Bioenergetics and Racehorse Ratings

Bob Wilkins

Overdee Press

Published in 2010 by Overdee Press

Printed by Rayross Print Factory, Liverpool, UK

Copyright © Bob Wilkins

First Edition

Bob Wilkins asserts the right under the Copyright, Designs and Patents Act 1988 to be identified as the author of this work.

All rights reserved. No part of this publication may be reproduced, stored in a retrieval system, or transmitted, in any form or by any means without the prior written consent of the publisher, nor be otherwise circulated in any form of binding or cover other than that in which it is published and without a similar condition being imposed on the subsequent purchaser.

ISBN 978-0-9564243-0-3

A CIP catalogue record for this book is available from the British Library

www.overdee.com

to

Chris

with love

Introduction

In horse racing each horse has to carry a specified weight, which includes that of the jockey and equipment. Any extra that is required is made up by lead weights carried in the saddle. For handicap races, the total weight is specified by the official handicapper, with the objective of equalising the chances of all the horses. For non-handicaps the weight is set out in the conditions specified for the race, older horses usually having to carry more weight than younger horses, according to a 'weight-for-age' scale.

The foundation of this system is the work of Admiral Rous in the nineteenth century. Rous studied the running of horses and the effect of weight, age and distance and he published *The Laws and Practice of Horse Racing*, which was the standard work for framing the conditions of races for many years. Although the detail of these recommendations has been changed many times since then, the basic system of handicapping by allocating different weights to different horses remains the same.

One of the key issues to be decided is the relationship between weight carried and the time taken to run a certain distance. Alternatively this can be considered as the relationship between weight and distance travelled in a certain time. In 2009, the website of the British Horseracing Authority (BHA) states that "in Flat races one length is typically reckoned to be worth three pounds in sprint races ...". This means that if horse A beats horse B by a length in such a race, it is considered to be 3 pounds superior to B. If in a future race A has to carry an additional 3 pounds, compared to B, their chances of winning are equalised.

Official ratings measured in pounds are published by the BHA and are used to determine the weights to be carried in handicap races. As well as the official ratings, there are many commercial organisations who publish their own ratings, and countless individuals who do their own 'private handicapping'.

Despite the fact that allocation of weights is one of the cornerstones of the racing industry, there are some who assert that weight has no effect on the running of a horse. This flies in the face of Newtonian mechanics and also everyday human experience. It takes longer to return home carrying two heavy shopping bags than it did to walk to the shop when the bags were empty.

In this book, a whole-body metabolic model of competitive running is described, which for the first time gives a physiological basis for the calculation of racehorse ratings. The model draws on published work in the fields of exercise physiology, veterinary science, biomechanics, physics, numerical analysis, and of course racing itself.

Specialised terms can be found in each of these fields, and as far as possible these are explained as they are introduced. In particular, the language of horse racing can be very confusing. For example the word 'handicapper' can mean the person who allocates the weights to be carried, or it can mean a horse who usually only runs in handicaps. Readers who may be unfamiliar with the sport can find an excellent introduction on the BHA website at *www.britishhorseracing.com*.

Whole-body metabolic models of running balance the energy supply from both anaerobic and aerobic sources with the energy required to accelerate the body, sustain running, and overcome air resistance. They are well established for modelling the running of human athletes.

The model described in this book is first developed and validated by application to the World Record times for sprinters and middle-distance male athletes, as these records stood after the World Athletics Championships in 2009. It is then extended to British horse racing on turf, to include the effects of weight, going, and difference between racecourses, for both flat and National Hunt racing.

Although whole-body metabolic models of horse racing have not been developed before, the parameters of the model which are calculated agree well with data that have been published in other works on equine exercise physiology. In addition the relationships between weight and time for standard conditions are in good agreement with the rules-of-thumb which are in common usage.

The result of the model is a Power Equation, which can be used to assess performance in a race in terms of a *power rating*. Two methods of assessing performance are examined in detail, based on race time, or on collateral form. An objective method of calculating collateral form ratings is also described.

Although power ratings can be converted to scales expressed in pounds, it is preferable to use power ratings because the Power Equation contains dependencies between weight, time, distance and going which are lost in the conversion. The calculation of power ratings is complex, and cannot practically be done using pencil, paper and pocket calculator, but nowadays, with a PC on almost every desk, this is not a problem.

The parameters which have been computed are applicable to racing on turf in Britain, but the model is easily adapted to racing on other surfaces and tracks.

This book is about the link between equine exercise physiology and racehorse ratings. The uses of these ratings are only briefly discussed, at the end of the book. However, because racing and betting are inextricably linked, a short cautionary note on betting is included in Appendix A.

Symbols are defined when they are first introduced in the text, but for reference purposes a list is included at the end of the book.

The help of the *Racing Post* is gratefully acknowledged, for making standard times available to the public on their excellent website *www.racingpost.com*. Thanks are also due to 'Supernap' of the Daily Herald, who stimulated my interest in racing as a teenager, and to Bill Ferguson, who rekindled that interest some 20 years later.

Bob Wilkins

Heswall, 2010

Table of Contents

Chapter 1. Energetics of Running in Athletics — 1

1.1 Sprints and distance events — 2

1.2 Energy sources — 4
    A . Anaerobic energy supply model — 6
    B . Aerobic energy supply model — 7

1.3 Energy expenditure — 8
    A . Energy cost of running, C_R — 8
    B . Air resistance — 10
    C . Kinetic energy — 11

1.4 Energy balance equation for a race — 12

1.5 Predictions of the optimised model — 16

1.6 Conclusion — 22

Chapter 2. Modelling of Running by Racehorses — 23

2.1 Energy sources — 26
    A . Anaerobic energy supply model — 26
    B . Aerobic energy supply model — 26

2.2 Energy expenditure — 27
    A . Energy cost of running, C_R — 27
    B . Air resistance — 28
    C . Kinetic energy — 29

2.3 Energy balance equation for a race — 29

2.4 Basic predictions of the model — 31

2.5 Further implications of the model — 33

		A . Effect of weight on time	*34*
		B . Effect of weight on distance	*36*
		C . Effect of going	*38*
		D . Going allowances	*39*
		E . Effect of pace	*41*
		F . Getting the trip	*43*
		G . Gain in altitude	*45*
		H . Other effects	*45*
	2.6	Jump racing	45
		A . Hurdles	*46*
		B . Chases	*47*
	2.7	Conclusion	50

Chapter 3. Assessment of Performance - Time-Based Ratings   51

3.1	Cost of running at a specific track C_R'	53
3.2	Beaten horses	54
3.3	Time-based ratings	55
3.4	Example	58
3.5	Conversion of power ratings to a scale given in pounds	64
3.6	Weight-for-age scales	67

Chapter 4. Assessment of Performance - Form-Based Ratings   68

4.1	Example	68
4.2	Conversion to pounds	75
4.3	Objective collateral form ratings	75
4.4	Sparsity-oriented programming	80

4.5	Application to power ratings	81
4.6	Large-scale application	84
4.7	Weight-for-age scales	85
	A . Effect of W on ratings	*86*
	B . Weight-for-age	*88*

Chapter 5. Allowance for Weight in Future Races   90

Chapter 6. Conclusion   94

References   95

Appendix A - a note on betting   *98*

Appendix B - numerical methods   *103*

Appendix C - Dickinson's model   *104*

Appendix D - list of principal symbols   *106*

Chapter 1. Energetics of Running in Athletics

This book describes a scientific study of the factors which govern how long it takes a racehorse to run a given distance, and in particular, the effect of the weight carried. While there is little published science on this topic, there is a wealth of information on human athletics. Top-class athletes run on level, good quality tracks, with race times accurately measured (to 0.01s) and athletes are usually happy to cooperate with researchers. The same cannot be said for horses, who in Britain run mostly on turf tracks with a myriad of course qualities and conformations, and who often have a mind of their own.

In this chapter a mathematical model of the running of elite male athletes will be described, which will be used as the basis for a model of the running of horses, in Chapter 2.

The model draws on research from two main disciplines, exercise physiology and biomechanics (McArdle *et al* 2005, Hay 1993).

Exercise physiology is concerned with the effect of exercise on the body, particularly the muscles, and the way in which the body produces the metabolic power to sustain exercise. It is often concerned with the detailed biochemical processes that occur within cells, but "whole-body" metabolic models have also been extensively researched (Morton 2006). *Biomechanics* is concerned with detailed analysis of the movement of the limbs and the rest of the body during activities such as walking, running and jumping, but generally pays little attention to the sources of energy and power which are needed to drive these activities.

The total metabolic power produced by the body has to be converted to mechanical power in order to produce physical motion. Energy is lost internally overcoming muscle viscosity and joint friction, sustaining isometric muscular contractions and driving those movements of the body and limbs which do not lead to a forward displacement of the runner's centre of gravity (Cavagna *et al* 1964). Relatively few researchers have studied the whole process, but it appears that the efficiency of conversion of metabolic power to mechanical power is less than 50%.

The rest of this chapter is concerned with a model of the running of elite male athletes, using outdoor World Record times as they stood in 2009. The model is based on energy balance at the metabolic, rather than the biomechanical level.

1.1 Sprints and distance events

Lloyd (1966) studied the record times for humans for a very wide range of distances and interpreted these in terms of the quantities of metabolic energy available at different times. Distance-time plots were interpreted as series of straight lines, the slopes of which decreased as running time increased.

Keller (1973) analysed world record times for sprints and middle distance events as they stood at that time. According to Keller, in a "sprint" event, the athlete runs "flat out" for the duration of the race, i.e. as fast as he possibly can. However, he cannot keep this up indefinitely, because his energy sources will eventually all be consumed and he will collapse with exhaustion. The distance at which this occurs may be called the "critical distance", which Keller calculated to be 291m. Thus 100m and 200m races would be classified as true sprints, while 400m would not.

If the athlete wishes to run further than the critical distance, his (constant) running speed will need to be reduced to conserve energy, to allow him to complete a longer trip. He will then continue running until he collapses with exhaustion when reaching the finishing line.

So with Keller's model, in "distance" events the runner's energy resources limit the best time which can be attained, while for "sprint" events it is the maximum propulsive force (or power) which is the key. More recent research (Busso *et al*, 2006) has shown that Keller's estimates of metabolic power and energy are much too high, but nevertheless the concepts described are very useful.

Table 1.1 shows the men's World Record times for the Olympic distances of 100m to 1500m as they stood after the World Athletics Championships in 2009. Also shown is the average speed. For distances greater than 200m the average speed drops significantly - the average speed for 1500m is about 30% lower than the speed for 100m.

Record times exist for other distances (e.g. 300m, 1000m) but these are not often run, and the records are not as representative as those for the Olympic distances.

D (m)	T (s)	V (m/s)
100	9.58	10.44
200	19.19	10.42
400	43.18	9.264
800	101.11	7.912
1500	206.00	7.282

Table 1.1 Men's Outdoor World Records, 2009

Figure 1.1 shows the World Records as a distance-time plot. There appears to be a change in slope at about 300m. If a straight line is drawn through the three upper points, it intersects a straight line through the two lower points at about 300m, which is similar to the value of critical distance proposed by Keller. The longest distance considered here (1500m) is five times higher than the critical distance.

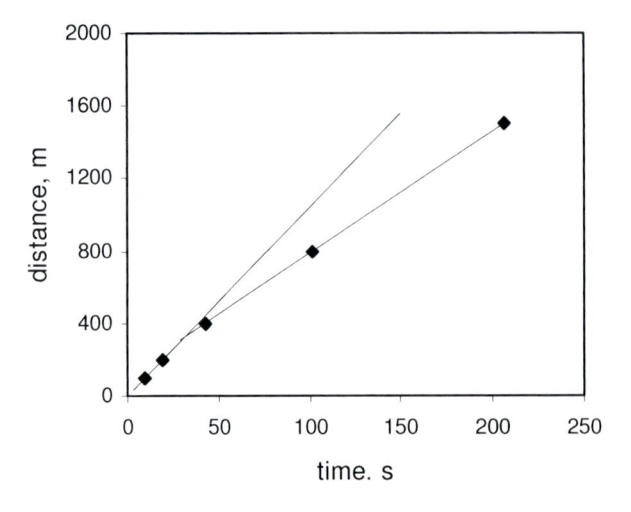

Figure 1.1 World record distance-time plot

In horse racing the range of distances run is much lower, and so it is not necessary to include the long-distance events (5000m, 10000m and the marathon) in our model.

1.2 Energy sources

There are several very similar models of running based on the whole-body metabolic energy balance (di Prampero *et al* 1993, Busso *et al* 2006, Peronnet and Thibault 1989). However the model used here is essentially that of Peronnet and Thibault.

Energy for muscle contraction is obtained from the chemical adenosine triphosphate (ATP) in muscle tissues. When ATP molecules split a large amount of energy is released. Replenishment of ATP when exercise is sustained comes via two "pathways", *anaerobic* and *aerobic*.

A small amount is stored in muscle tissue and is available immediately after exercise begins. However, after a few seconds this is replenished from other energy stores in the muscles, blood glucose, and glycogen from the liver, but the total amount of energy available for a prolonged bout of exercise is limited. These processes do not require oxygen and are termed *anaerobic*.

The second mechanism for replenishing ATP uses blood-borne oxygen via the respiratory and circulatory system (the *aerobic* pathway).

The biochemistry of these processes is complex (McArdle 2005) and beyond the scope of this book.

Aerobic power derives from the oxygen absorbed via the lungs into the bloodstream and can be quite accurately measured. Measurement is a routine procedure for athletes and others who engage in physical activity, for example, firefighters whose fitness needs to be checked. The rate of oxygen absorption is conventionally denoted $\dot{V}O_2$ and is usually expressed in ml/min/kg of body mass. (The dot over the V in this composite symbol denotes that this is a rate quantity).

After running begins, $\dot{V}O_2$ increases and reaches a stable steady-state value. If the demand for power is *submaximal* (below the capacity of the athlete), the aerobic power produced meets the demand, and the athlete can, in theory, continue to run indefinitely. However, when running speed increases, there comes a point when $\dot{V}O_2$ reaches the limit that particular athlete can convert ($\dot{V}O_2 max$).

If exercise is *supramaximal*, aerobic supplies are not sufficient to sustain the activity indefinitely, and anaerobic stores are then used, which will subsequently run down until the athlete is exhausted.

One millilitre of oxygen absorbed via the lungs corresponds to 20.9J of metabolic power (di Prampero *et al* 1993). It is common to normalise metabolic energy and power by dividing by the athlete's body mass to obtain values expressed in J/kg or W/kg.

A. Anaerobic energy supply model

The total anaerobic energy store E_0 for an elite male athlete is approximately 1400-1600 J/kg (Peronnet and Thibault 1989). However this is not all immediately available; for supramaximal work the available anaerobic metabolic energy supply grows as shown in Figure 1.2.

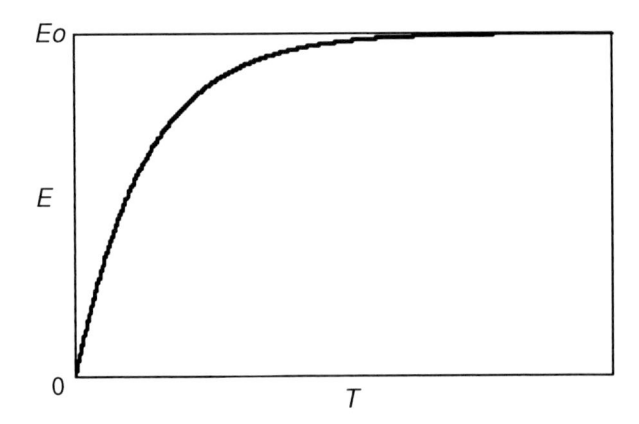

Figure 1.2 Available anaerobic energy as a function of time

The energy shown on the vertical axis is the amount *which is available for a race of duration T*. It can be approximated by the exponential growth model of equation (1).

$$E' = E_0 (1 - e^{-T/T_E}) \tag{1}$$

where T_E is the time-constant governing the growth and is not known precisely. Lloyd (1966) estimated T_E as 25s, while others have used values from 10s to 30s (di Prampero *et al* 1993, Peronnet and Thibault 1989).

B. Aerobic energy supply model

For supramaximal exercise the *instantaneous* aerobic metabolic power grows as an exponential function of time, according to

$$p = P_m (1 - e^{-t/T_P}) \tag{2}$$

where P_m is the maximum aerobic power (derived from $\dot{V}O_2 max$) and T_P is the associated time-constant. For elite male athletes P_m is about 28W/kg (~ 80mlO$_2$/min/kg). However again T_P is not known precisely, and values from 10s to 30s have been used by various authors (Peronnet and Thibault 1989).

The total energy delivered aerobically is obtained by integrating (2) over the duration of the race.

$$E'' = \int_0^T p\, dt = P_m \int_0^T (1 - e^{-t/T_P})\, dt$$

which gives

$$E'' = P_m \left[T - T_P (1 - e^{-T/T_P}) \right] \tag{3}$$

The average aerobic power over the race duration is obtained by dividing by T, to get

$$P = P_m \left[1 - \frac{T_P}{T}(1 - e^{-T/T_P}) \right] \tag{4}$$

P is shown as a function of T in Figure 1.3.

Maximum aerobic power can only be sustained for about 7 minutes, after which it begins to fall gradually (Busso *et al* 2006). This effect needs to be taken into account if long-distance races such as the 5000m, 10000m and marathon are to be modelled, but it does not need to be considered here.

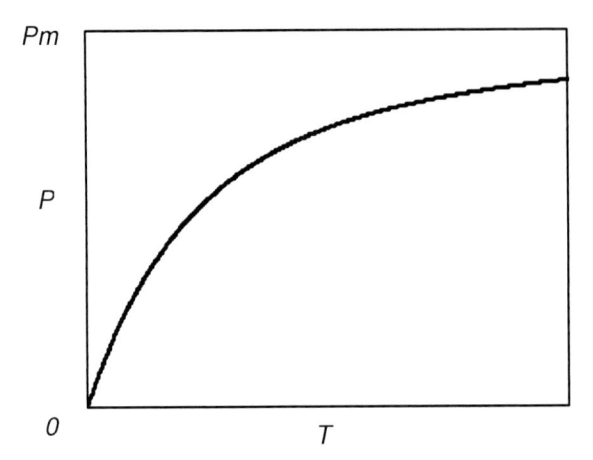

Figure 1.3 Average aerobic power supplied as a function of duration.

P_m is the value *above resting*. When the body is at rest and performing no external work a small amount of power is needed to sustain life. This is the *basal metabolic rate*, and is generally taken to be about 3.5 mlO$_2$/min/kg (~ 1.2 W/kg). For our model of running, this is taken as a baseline, and only the additional power needed to sustain exercise is considered.

1.3   Energy expenditure

The energy sources described in the previous section are used in three principal ways.

A . *Energy cost of running, C_R*

The so-called energy cost of running C_R (J/kg/m) is the energy needed to move each kilogram of body mass through a distance of 1 metre. So for a distance D the energy required in J/kg is

$$E_R = C_R D \qquad (5)$$

Dividing by the race time T gives the average power needed in W/kg.

$$P_R = C_R \frac{D}{T} \tag{6}$$

C_R has been extensively measured for humans and animals over many years using tests on a treadmill. In these tests the subject runs on a horizontal treadmill for several minutes at different submaximal speeds. At each speed the steady-state rate of oxygen intake $\dot{V}O_2$ is measured. This is then converted to power in W/kg (using 20.9J per mlO_2) and the slope of the graph of power versus speed (D/T) gives the value of C_R.

C_R is independent of speed and is of the order of 3.5-4.0 J/kg/m. di Prampero *et al* (1993) give an average, from values reported in the literature, of 3.86 J/kg/m. Equation (6) gives the average metabolic power to sustain running at a given speed. Multiplying by the efficiency gives the mechanical power delivered at "ground level" which is used to drive the runner forward.

At the ground level there are rapid fluctuations in mechanical power, dictated by the runner's stride. The stride consists of three phases (a) the *push*, during which the rear leg pushes the runner's centre of gravity upwards and forwards against the effect of gravity, (b) the *flight*, during which neither foot touches the ground, and (c) the *landing* when the leading leg hits the ground, producing slight deceleration and a return of some energy because of muscle elasticity, before moving forward for the next stride. For a typical elite athlete the flight lasts for about 45% of the total stride (Hay 1993).

The pulsating nature of the mechanical power has been recorded by several researchers (Cavagna *et al* 1964, Cavagna *et al* 1971, di Prampero *et al* 2005). However whole-body metabolic processes cannot respond quickly enough to the stride pattern, and an average value of

C_R is sufficient to represent the power requirement.

(Note that if the treadmill speed is increased further, there comes a point when $\dot{V}O_2 max$ is reached and no further increase in $\dot{V}O_2$ is possible. This is the method used for determining $\dot{V}O_2 max$ and hence P_m).

B. Air resistance

When running on a treadmill in still air there is no relative motion between the runner and the mass of the surrounding air. However, when running at a speed v on a track in still air there is a viscous drag force given by the well-known relationship from fluid mechanics

$$F = \frac{1}{2} \rho A C_D v^2 \qquad (7)$$

where

$\rho =$ air density, (approximately 1.2 kg /m^3 at sea level and 20C).

$A =$ projected area (normal to the motion, approximately 0.45m^2 for a male athlete) (Davies 1980).

$C_D =$ dimensionless drag coefficient.

Dividing by the body mass of the runner W gives f in N/kg, as

$$f = k v^2 \qquad (8)$$

Where $k = 0.5 \rho A C_D / W$.

For a typical male athlete of 70kg body mass, $k \sim$ 0.5 x 1.2 x 0.45 x 1.0 / 70 = 0.0038, if the drag coefficient is assumed to be 1.0.

Hill (1927), Pugh (1971) and Davies (1980) have carried out experiments in wind tunnels with models and real athletes. These results confirm the approximate value of k, although most of the experiments show that the drag coefficient C_D is less than 1, of the order of 0.8. For the 100m sprint the projected area of the runner is very low at the start of the race, and then increases as the runner rises from the crouched position to the fully upright position. Mureika (2001) suggests that for these conditions the drag coefficient is of the order of 0.5.

Using (8), for a race of distance D run in a time T the average energy required per kilogram to overcome air resistance is (*force × distance*)

$$E_A = k \left(\frac{D}{T}\right)^2 D \qquad (9)$$

while the average power required is obtained by dividing by T, as

$$P_A = k \left(\frac{D}{T}\right)^3 \qquad (10)$$

The estimate of k given previously refers to the mechanical requirement. At the *metabolic* level a larger amount of power will be required, as the conversion of metabolic power to mechanical power is not 100% efficient. The results of Pugh (1971) and Davies (1980) show that to estimate the metabolic power to overcome air resistance, k needs to be 0.0057-0.0089 and 0.0093-0.0097 respectively. This implies that the efficiency of conversion in overcoming air resistance is of the order of 35%.

In 1984 di Prampero used $k = 0.01$ to estimate the metabolic power requirement, and this value has subsequently been adopted by others (Busso et al 2006, Peronnet and Thibault 1989).

C. Kinetic energy

An object of mass m moving with a velocity v has a stored kinetic energy of $mv^2/2$ and the runner has to supply this during the acceleration period at the beginning of the race. di Prampero and others assume that at the beginning of a race the efficiency of conversion of metabolic energy to kinetic energy is only 25%, as there is no recovery of elastic energy at this stage (di Prampero 1993). So the metabolic energy required is $2mv^2$. Dividing by the body mass and putting $v = (D/T)$ then gives

$$E_K = 2\left(\frac{D}{T}\right)^2 \qquad (11)$$

and the corresponding average power is

$$P_K = 2\left(\frac{D}{T}\right)^2 \cdot \frac{1}{T} \qquad (12)$$

1.4 Energy balance equation for a race

The time required to run a distance D to a given standard, such as a World Record, can be obtained by balancing the energy available and the energy needed, as given by equation (13) below.

$$\underbrace{E_0(1-e^{-T/T_E})}_{\text{anaerobic supply}} + \underbrace{P_m\left[T - T_P(1-e^{-T/T_P})\right]}_{\text{aerobic supply}} = \underbrace{C_R D}_{\substack{\text{cost of} \\ \text{locomotion}}} + \underbrace{k\left(\frac{D}{T}\right)^2 D}_{\substack{\text{air} \\ \text{resistance}}} + \underbrace{2\left(\frac{D}{T}\right)^2}_{\substack{\text{kinetic} \\ \text{energy}}}$$

... (13)

All terms in (13) have the dimensions of J/kg. Alternatively, by dividing throughout by T we obtain the average power balance equation (14).

$$\frac{E_0}{T}(1-e^{-T/T_E}) + P_m\left[1-\frac{T_P}{T}(1-e^{-T/T_P})\right] = C_R\frac{D}{T} + k\left(\frac{D}{T}\right)^3 + 2\left(\frac{D}{T}\right)^2\frac{1}{T}$$

... (14)

The model is characterised by the six parameters $[E_0, T_E, P_m, T_P, C_R, k]$. If these are known the time required to run a distance D can be obtained by solving (13), or (14) for T. An explicit solution cannot be derived because the equations are complicated and non-linear because of the exponential terms. However a solution can be obtained very easily using numerical methods, as the equations are well-behaved. Figure 1.4 shows the energy supply (left-hand side) of equation (14) and the usage (right-hand side), as a function of T for $D=800$m. The solution is the point where the two curves cross, and can be computed using the following very simple numerical method.

First, assume that the solution is "bracketed" (Press *et al* 2007) between a low value T_L and a high value T_H. For our World Record data we can take $T_L = 0.1$s and $T_H = 100,000$s, since all solutions lie within this range. Then ...

1. Calculate the mid-value $T_M = 0.5(T_L + T_H)$

2. Calculate the power usage and supply at T_M

3. If usage at T_M equals supply at T_M (within a desired accuracy) - done

4. If usage at T_M exceeds supply at T_M then ...

>(T_M lies to the left of the solution point)
>
>reset T_L to T_M (move left bracket up)
>
>go to step 1

5. If usage at T_M is less than supply at T_M then ...

(T_M lies to the right of the solution point)

reset T_R to T_M (move right bracket down)

go to step 1

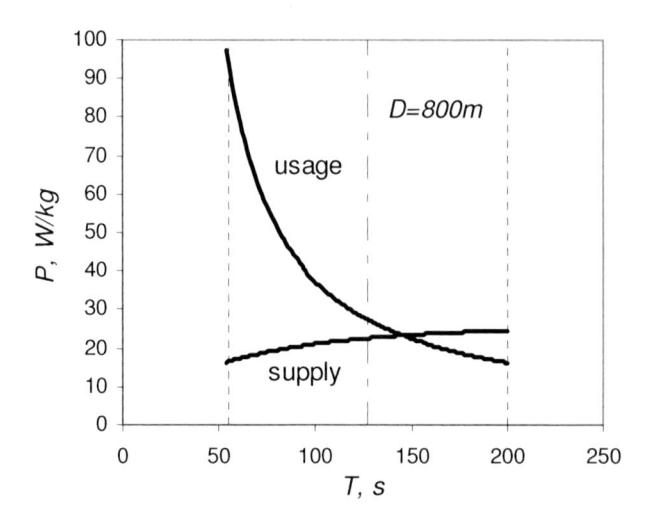

Figure 1.4 Solution by method of bisection

Although mathematically the solution is simply the point where the two curves cross, which can be obtained very quickly using the simple algorithm given, it is useful to inspect Figure 1.4 more closely, to reinforce the concepts of the model.

For times to the left of the intersection point of the curves, the energy supply available is less than that required to run the distance D, and the runner could not possibly complete the race in that time. On the other hand, for times to the right of the intersection point, the energy supply available is greater than required, so the runner could actually complete the race in a shorter time i.e. it would not be a World Record.

It remains to determine the values of the model parameters which are needed to match the World Record data given in Table 1.1. Since the cost of running coefficient C_R is well documented in the literature, it was taken to be the value used by di Prampero, 3.86 J/kg/m. This leaves the parameters [E_0, T_E, P_m, T_P, k] to be found. In principle since there are 5 parameters to be found and 5 sets of data in Table 1.1, an exact solution is possible, but because of the non-linearity of the equations a numerical method is again necessary. The following simple brute-force method was used.

1. Set initial estimates (from the literature) of [E_0, T_E, P_m, T_P, k] = [1500, 20, 28, 20, 0.01]
2. Calculate the times predicted by the model for each of the 5 distances by solving (14)
3. Calculate the normalised sum of the squares of the errors
$$SSE = \sum_{1}^{5}[(T_{wr} - T_{model})/T_{wr}]^2$$
4. Vary each parameter, in turn, by a small *random* fluctuation (positive or negative)
5. Recalculate *SSE* and if a lower *SSE* is obtained, accept the new value of the parameter. Otherwise, retain the old value
6. Go to step 3 and repeat until no further reduction in *SSE* is obtained.

After a few hundred iterations the value of *SSE* stabilises, and the final model parameters shown in Table 1.2 are obtained.

E_0	1541.9	J/kg
T_E	16.66	s
P_m	28.69	W/kg
T_P	24.12	s
k	0.010871	Ws^3/kg-m^3

Table 1.2 Optimised parameters for men's World Records

These parameters give an exact match to the World Record times given in Table 1.1 (subject to computational precision) and are within the ranges found within the literature. The value of P_m corresponds to $\dot{V}O_2max$ = 82.4 mlO_2min^{-1}. Note that the ratio E_0/P_m has the dimensions of time, with a value of 1541.9 / 28.69 = 53.7 s. It is the time it would take the maximum aerobic supply to deliver an energy equal to the anaerobic store.

1.5   Predictions of the optimised model

Figure 1.5 shows the relationship between distance and the predicted average time to run that distance, for distances up to 1500m. Also shown are World Records from Table 1.1. The curve clearly has two slopes. One for short times ("sprints") and a lower slope for longer times ("middle-distances"). However there is a region in between where the slope changes gradually from the first to the second. There is no sharp delineation of critical distance as suggested by Keller, but a transition zone between sprints and middle-distance races

Figure 1.6 shows the average total metabolic power (anaerobic plus aerobic) as a function of race time. This agrees well with the curve published by di Prampero et al (2005), who showed that elite athletes are capable of producing around 100 W/kg at the beginning of exercise.

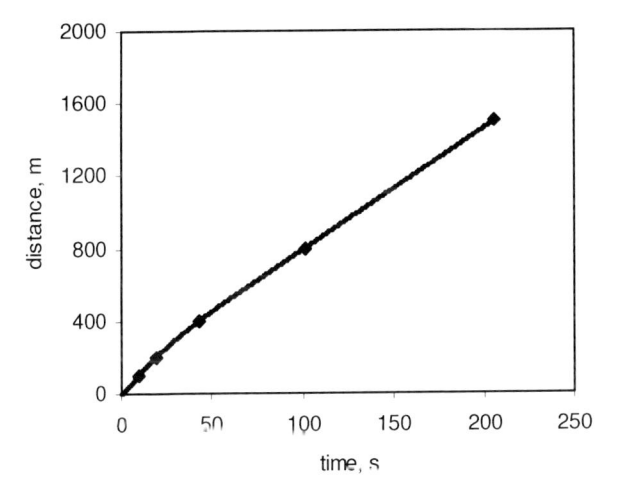

Figure 1.5 Distance-average time curve predicted by model

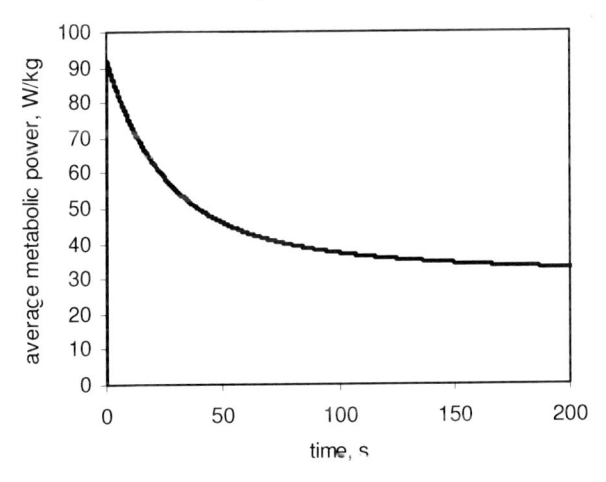

Figure 1.6 Total metabolic power supply

Figure 1.7 shows the split between anaerobic and aerobic power as a function of race duration. For short times the bulk of the energy comes from anaerobic supplies, but for long times aerobic energy dominates. Duffield and Dawson (2003) reviewed work on the anaerobic/aerobic split, and obtained results similar to those shown.

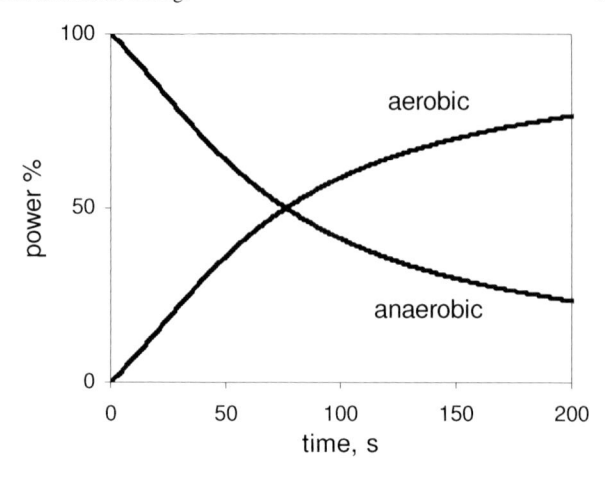

Figure 1.7 Percentage share of aerobic and anaerobic power

Figure 1.8 shows the percentage split between the three mechanisms of energy expenditure. For very short times most of the energy is converted to kinetic energy (the initial acceleration period) but for long times the kinetic energy term becomes a very small percentage of the total. The effect of air resistance rises and reaches a maximum for the longer sprints and then declines because the runner's speed reduces. For longer distances the "locomotion" term calculated from C_R is by far the most important.

Figure 1.9 shows the average speed predicted by the model as a function of race duration. The World Record points are also shown. For most of the recent history of world records, the speed for the 200m record has been slightly higher than that for 100m, because of the effect of the start. However in 2008-9 both these records were broken by Usain Bolt, and unusually, his speed over the 200m was slower than that achieved during his 100m gold medal run at the Beijing Olympics.

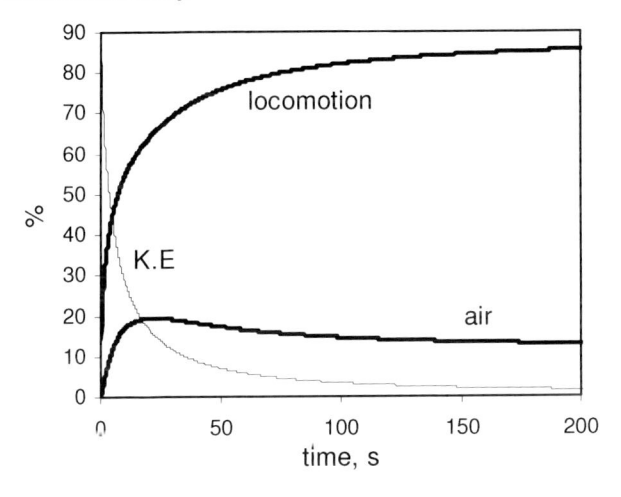

Figure 1.8 Percentage shares of power usage

As previously discussed, Figure 1.9 shows a transition zone from sprints to middle distance events rather than a sharp discontinuity.

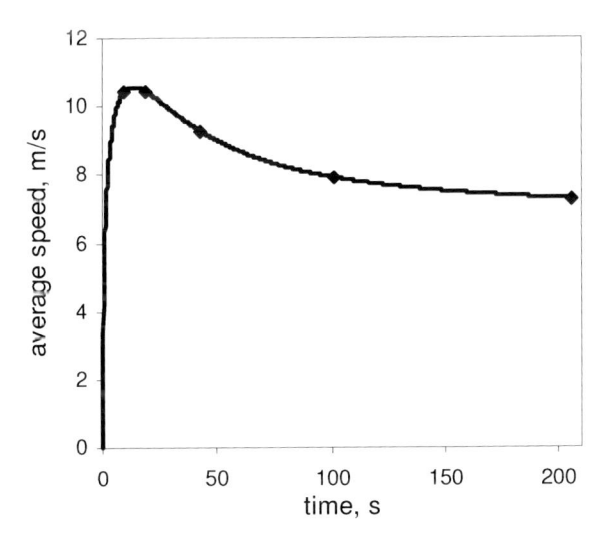

Figure 1.9 Predicted average speed as a function of race time

Overall, the model gives excellent agreement with World Record times and the physiological and physical processes which affect them. Of course World Record times are a snapshot of a particular time in athletic history. When one of these records gets broken, the model parameters will need to be re-computed. However these records only change by a very small percentage. The required adjustment to the model parameters will be small, and the model will still be a good representation of the running of elite athletes - the principles will not change.

The model makes no specific assumptions about any variation of speed during the race. It just requires that supramaximal effort is applied, so that the energy sources become available according to the left-hand side of (13) or (14). On the right-hand side, *average* values of the terms over the race duration are used.

Sprints

The model gives remarkably good results for sprints, given that it has been traditionally assumed that sprints are limited by maximum force rather than metabolic energy. Figure 1.10 shows the predicted average speed as a function of duration for times less than 10s. The 100m World Record is also shown.

The curve shown in Figure 1.10 is close to that which has been measured for elite sprinters (Cavagna *et al* 1971, di Prampero *et al* 2005, Mureika 2001). This is remarkable but somewhat fortuitous, as the model does not explicitly contain any representation of the dynamics of acceleration of the runner immediately after the start, although the kinetic energy term in the model deals with this implicitly, and is dominant for very short times - as occurs in reality.

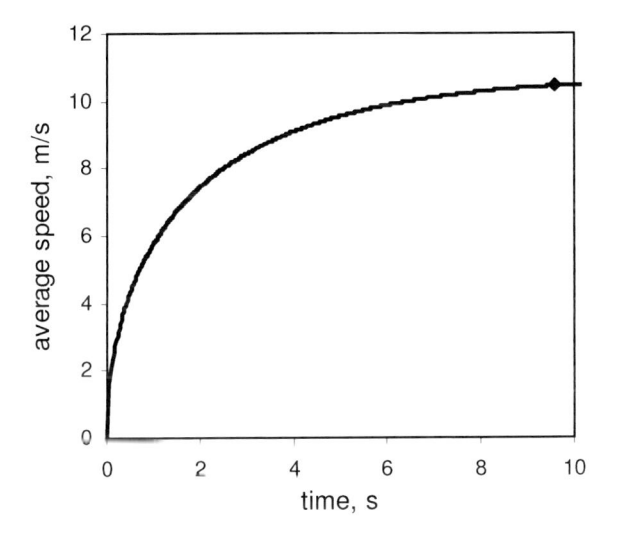

Figure 1.10 Predicted average speed for sprints

Mureika's model represents these processes in great detail, including the change in posture of the athlete after leaving the starting blocks. He used appropriately chosen mathematical functions to represent the accelerating forces, and fitted the results to the velocity-time profile of a top athlete, measured at 10m intervals over a 100m race. There was no modelling of metabolic processes, just upper physical limits imposed on performance parameters. Mureika showed that top 100m sprinters reach peak speed after about 60m, and thereafter the speed actually falls slightly, so the extreme physical effort appears to cause some type of muscular exhaustion. This means that Keller's idea that the athlete can sustain his maximum speed until the critical distance is reached is not strictly correct.

Another factor not considered by the metabolic model is that the 200m race includes a bend, which slows the runner down a little. There is no doubt that if the 200m were run over a straight track, the World Record would be lower (Mureika 1997).

It is simple to allow for the effect of a tail wind in the model. The energy required to overcome air resistance changes to

$$E_A = k\left(\frac{D}{T} - V_W\right)^2 D \qquad (15)$$

where V_W is the speed of the tail wind. A sprint record cannot be ratified if the tail wind speed exceeds 2m/s. It is generally accepted that a tail wind of this magnitude reduces the 100m time by 0.1-0.12s (Mureika 2001). However using (15) with the model developed here gives a reduction of 0.29s. This error is mainly due to the fact that in a real sprint the drag coefficient and projected area are reduced for the initial part of the race - this would require a time-dependent air-resistance term to be included in the energy balance model.

A further complication with sprints is the *reaction time*. Times are measured electronically from the firing of a starting pistol, and the runner's reaction time is measured. A reaction time of less than 0.1s is considered impossible, and such a case is deemed to be a false start.

Dynamic modelling of these and other secondary processes involved in true sprint running is very complicated, and linking these to physiological processes represents a real challenge.

1.6   Conclusion

In this chapter a model of running based on the work of Peronnet and Thibault has been described in detail. In the next chapter this model will be applied to the running of racehorses for which the basic physiological and physical laws still apply, but some extensions and modifications will be needed.

Chapter 2. Modelling of Running by Racehorses

In athletics a runner's fastest time over a particular distance is the most important statistic, which indicates his ability relative to other competitors - the World Record times given in the previous chapter were recorded by the greatest athletes of the time. Where there are heats in championships, the heat winners plus some of the fastest losers go through to the next round. This illustrates the dominant role of race times in assessing athletic performances.

With horseracing on turf in Britain, race times are much more difficult to interpret, so much so that some followers of the sport ignore them completely. While athletes run on a flat track with a consistent surface, horses run on racecourses with many different conformations, with widely variable ground conditions.

According to the particular racecourse, some distances are run on the level, some uphill, some downhill, some undulating. Courses may include left-hand turns, right-hand turns, even figures-of-eight. In some cases the turns are very tight, while in others they are gradual. Although race times are officially published, they are not measured with the same degree of accuracy as with athletics, and neither are the distances. The raw race time at a particular course has to be considered in conjunction with the course conformation.

An even greater problem is the effect of the ground condition, or "going". Typical descriptions of the going are heavy, soft, yielding, good, firm, or hard, according to the nature of the turf and the degree of moisture retained in it. Heavy going gives slow races times, and refers to very muddy conditions after prolonged rainfall; anything wetter than this usually results in racing being abandoned. Hard going gives fast race times, and usually occurs after drought. It can potentially cause leg injury for some racehorses, but this is alleviated by watering the course.

The effect of the going on race times is enormous. A one mile race is normally completed in about 100s, but the influence of the going can easily cause this to vary by ± 5%, corresponding to a variation in distance of ± 80m.

Record times are of little use. While the athletics records given previously reflect the performance of truly great champions, record times at racecourses are invariably produced when the going is very firm and the race just happens to have been run at a fast pace. At top-class courses such as Ascot, the names of great racehorses can occasionally be found among the record-holders, but this is not true for the dozens of courses at which the minor meetings are held.

A further difference between athletics and horseracing is that horses may carry different weights (rider + additional weights carried in the saddle) each time they run, and this also has an effect on the race time.

Many horse races are run tactically. At the worst, some races see riders dawdling, unwilling to take the lead, resulting in very slow times which are meaningless. On the other hand, races which, according to observers, are run competitively throughout are referred to as being "truly-run".

The effect of different course conformations can be dealt with to a large extent by the use of *standard times*. Various organisations publish standard times for each distance run at each racecourse. They are unofficial and different organisations may evaluate them differently. A typical definition is that the standard time is the "time required to complete the distance when carrying 9st (flat racing) on good going in a truly-run race".

Compilation of a list of standard times is a daunting task that requires the study of races at all distances at all racetracks over recent decades. Most experts use subjective judgements to decide what the standard should be. Others use the statistical mean or median of times recorded over many decades. However, mean or median times are not good measures, as the distribution of times is skewed towards the longer times, because so many races are slowly run. Furthermore the median times at top-class courses such as Ascot and Sandown tend to be relatively lower, compared to courses which only have low-quality racing. This reflects the class of the horses which run at these courses, rather than the inherent challenge of the track itself.

To evaluate the parameters which fit a Peronnet-Thibault-type model for British flat racing, the standard times published by the *Racing Post* have been used (see *www.racingpost.com*). These cover 339 sets of distance-time data for 35 racecourses, with distances run from 5f to almost 22f.

[Note : Distances in British horseracing are given in miles, furlongs, and yards, and these units will also be used for the display of results, although they are converted to metres for the calculations. (1mile = 1609.3m; 1f = 201.16m; 1yd = 0.9144m).]

Lloyd (1966) observed that horses run about twice as fast as men, and that they "keep it up much better". He plotted some record times as they stood in 1965 against distance, and noted a change in slope at about 66s. This corresponds to a distance of about 6f and implies that the critical distance for racehorses is about 6f. Although the physiology of exercise in the horse and other animals is similar to that in man (Roberts *et al* 1998), there are obvious differences of scale, and the parameters of the model for horseracing need to be determined.

The effect of the weight carried is represented by two additional variables

> W = weight of the horse
>
> w = weight carried

Weights are usually expressed in stones and pounds (1st = 14 lb). Note that for these two variables the common term "weight" is used, rather than the more correct "mass". For flat racing the weight of the standard horse is assumed to be 1100lb (500kg). (See Chapter 4 for details concerning the choice of this value).

2.1 Energy sources

A. Anaerobic energy supply model

The value of E_0 for horses is not well documented, but an initial estimate based upon Lloyd's observation would be double the value for male athletes, i.e. about 3000 J/kg. The total anaerobic energy available is WE_0. No data is available on the value of the anaerobic time constant T_F. An initial estimate of T_E is 20s.

B. Aerobic energy supply model

Potard et al (1998), McCutcheon et al (1999) and many others have measured $\dot{V}O_2 max$ for thoroughbred racehorses using treadmill experiments. Values in the range of about 150-180 mlO$_2$/min/kg and as high as 198 mlO$_2$/min/kg have been reported. Rose et al (1990) found that after being out of training for 6 weeks, one horse's $\dot{V}O_2 max$ fell to 154 mlO$_2$/min/kg, but after 8 weeks of training it rose to 192 mlO$_2$/min/kg. This is over twice the value for male athletes. Owners are understandably reluctant to allow their best horses to be subject to treadmill experiments.

Relatively few races are completed in the standard time or better, which represents a high level of performance, so a reasonable initial estimate of $\dot{V}O_2 max$ for the standard horse is 200mlO$_2$/min/kg. This corresponds to $P_m = 69.666$ W/kg. The total aerobic power available is WP_m. Again P_m will be taken to be the value above resting, which for horses is about 4.9 ml O$_2$/min/kg (~ 1.7 W/kg) (Potard *et al* 1998).

No precise value is documented for the aerobic time constant T_P, but several researchers report that for thoroughbreds, the heart rate and pulmonary exchanges increase more rapidly than for humans, and stabilise after about 60s. An initial estimate of T_P is 15s, for which $\dot{V}O_2$ would increase to 98.2% of $\dot{V}O_2 max$ after 60s from equation (2).

Comparing the racehorse with other species, it appears to be supreme in its ability to produce very high values of aerobic power and to sustain this for long periods - the outcome of hundreds of years of breeding.

2.2 Energy expenditure

A . *Energy cost of running, C_R*

The energy cost of moving one kg of body mass through a distance of one metre varies remarkably little between different species of running animals (Roberts *et al* 1998). However C_R decreases slightly as body mass increases. Potard *et al* published data for thoroughbred horses running on a level treadmill, which corresponded to $C_R = 2.87$ J/kg/m - lower than for male athletes. Many other researchers have performed similar tests, but usually with the treadmill set at an uphill slope (McCutcheon *et al* 1999), in order to reach $\dot{V}O_2 max$ more quickly, but this gives a higher value of C_R.

A treadmill has a hard surface relative to turf so on good going on flat turf C_R should be greater than 2.87. Also while the treadmill surface is flat, so-called "flat" racecourses hardly ever meet that description in reality. There are usually uphill and downhill sections, and bends, and the energy saved when running downhill is lower than that expended when running the same gradient uphill. A reasonable initial estimate of C_R is 4.0.

Since the stride begins with an upward and forward push, the total cost of running is $(W+w)C_R$ J/m, since the weight to be lifted and propelled forwards is that of the horse plus the weight carried.

Having four legs rather than two is not as much of an advantage *per se* as may be thought at first (Roberts *et al* 1998). More important is the fact that when running, an elite racehorse is only airborne for about 28% of the stride, so it spends more of its time doing useful work.

B. Air resistance

The racehorse with a crouched rider is a more streamlined object than an upright human, so the drag coefficient should be lower - say about 0.5. Head-on photographs show that the projected area is about 4 times that of a human, so the drag force due to air resistance should be about twice (i.e. 0.5 x 4) that for a human athlete. However since the coefficient k is normalised to body weight, we need to divide by the weight of the horse (1100lb [500 kg]) rather than the 70kg which was assumed in section *1.3.B* for a male athlete.

So for horses,

$$k \sim 0.01 \times 2 \times (70/500) = 0.0028$$

which can be used as an initial estimate. Per kilogram this is less than one-third of the value for athletes, which reflects the fact that air resistance has less relative effect on heavy objects than it does on light ones - a feather falls more slowly than a cannonball. However this does not mean that air resistance is necessarily less important for horses than athletes, because horses run about twice as fast and the drag force due to air resistance depends on the square of the speed.

For a race of distance D the *total* energy required to overcome air resistance is

$$E_{Atot} = kW\left(\frac{D}{T}\right)^2 D \qquad (16)$$

Note that in (16) the weight of the horse W has been used rather than $(W+w)$. This is to reflect that for a standard horse and rider the air resistance is not affected by the weight carried. The weight carried is varied by adding lead weights to pockets in the saddle, and does not affect the projected area.

C. Kinetic energy

The model used here is the same as for the human athlete, except that the total mass to be accelerated is $(W+w)$, requiring a metabolic energy of $2(W+w)(D/T)^2$.

2.3 Energy balance equation for a race

Balancing the energy sources in section *2.1* with the usage terms in section *2.2* gives

$$WE_0(1-e^{-T/T_E}) + WP_m\left[T-T_P(1-e^{-T/T_P})\right] = (W+w)C_R D + kW\left(\frac{D}{T}\right)^2 D + 2(W+w)\left(\frac{D}{T}\right)^2$$

$$\ldots (17)$$

and dividing throughout by W gives

$$E_0(1-e^{-T/T_E}) + P_m\left[T-T_P(1-e^{-T/T_P})\right] = \left(1+\frac{w}{W}\right)C_R D + k\left(\frac{D}{T}\right)^2 D + 2\left(1+\frac{w}{W}\right)\left(\frac{D}{T}\right)^2$$

$$\ldots (18)$$

The left-hand side of (18) gives the anaerobic and aerobic energy supplies per kg of the horse's body weight, and is identical in form to (13), for athletes. However, on the right-hand side the locomotion and kinetic energy terms are increased by the factor $(1+ w/W)$, which varies as the weight carried changes. The air-resistance term does not vary with w.

The corresponding power balance equation is

$$\frac{E_0}{T}(1-e^{-T/T_E}) + P_m\left[1-\frac{T_P}{T}(1-e^{-T/T_P})\right] = \left(1+\frac{w}{W}\right)C_R\frac{D}{T} + k\left(\frac{D}{T}\right)^3 + 2\left(1+\frac{w}{W}\right)\left(\frac{D}{T}\right)^2\frac{1}{T}$$

... (19)

The parameters of equation (19) were fitted to the standard time data using the same basic method as was used for male athletes in Chapter 1, using W=1100lb and $w = w_S$ = 9st = 126lb. However since C_R is not known precisely, it was added as a parameter to be determined by the fitting process. So the initial estimates of the model parameters were $[E_0, T_E, P_m, T_P, C_R, k]$ = [3000, 20, 69.666, 15, 4, 0.0028].

The set of (D,T) values represented by the 339 standard times contains considerable scatter, compared with the World Record athletic times used in Chapter 1. For this problem the stochastic optimisation process was improved, by adding a simulated annealing component (Press *et al* 2007), which gave faster convergence and a lower sum-of-squares of errors.

The model parameters which give the best fit to the standard times are given in Table 2.1.

E_0	2615.3	J/kg
T_E	18.500	s
P_m	68.296	W/kg
T_P	15.054	s
C_R	3.728	J/kg/m
k	0.002273	Ws³/kg-m³

Table 2.1 Parameters for standard model

The value of P_m corresponds to $\dot{V}O_2 max$ = 196.07 mlO₂/min/kg.

2.4   Basic predictions of the model

Figure 2.1 shows the distance- time characteristic predicted by equation (19), using the standard model parameters given in Table 2.1. The "cloud" of data points in the background consists of the 339 sets of *(D,T)* standard time data. The standard error for the fit is 0.0184.

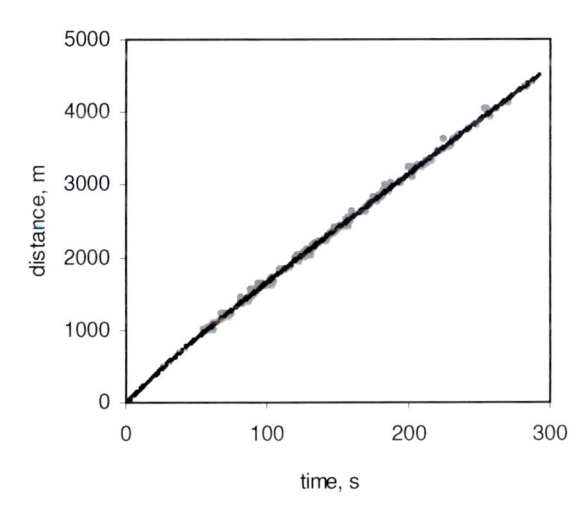

Figure 2.1 Distance versus race time

A bend in the curve is evident covering the transition from "sprint" to "distance" events, but the change in slope is less marked than for human athletes.

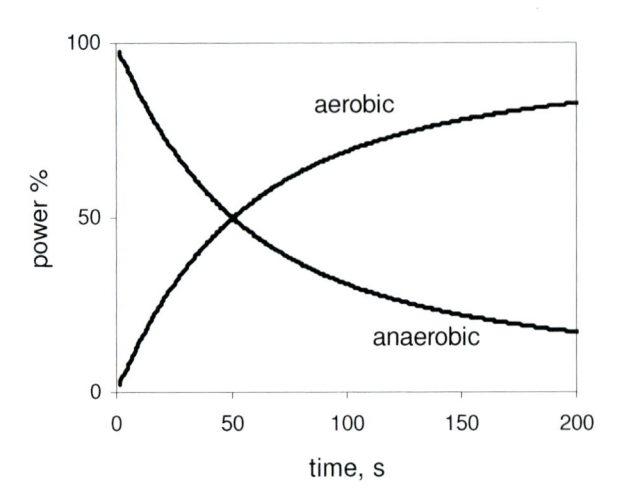

Figure 2.2 Percentage share of aerobic and anaerobic power

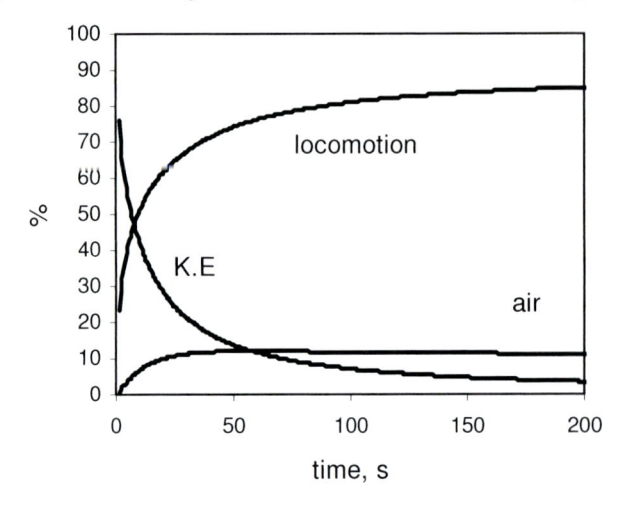

Figure 2.3 Percentage share of power usage

Figure 2.2 shows the split between anaerobic and aerobic power as a function of race duration. Aerobic power becomes larger than anaerobic power for race durations greater than about 50s. This is much shorter than for humans, for whom the corresponding time is about 75s. This is principally caused by the much shorter time constant T_P for horses.

The percentage shares of power for locomotion, air resistance, and kinetic energy shown in Figure 2.3 are similar to those given previously for humans.

The predicted average speed of a race, shown in Figure 2.4 reaches a peak at about 30s, and then declines, but much more slowly than for humans.

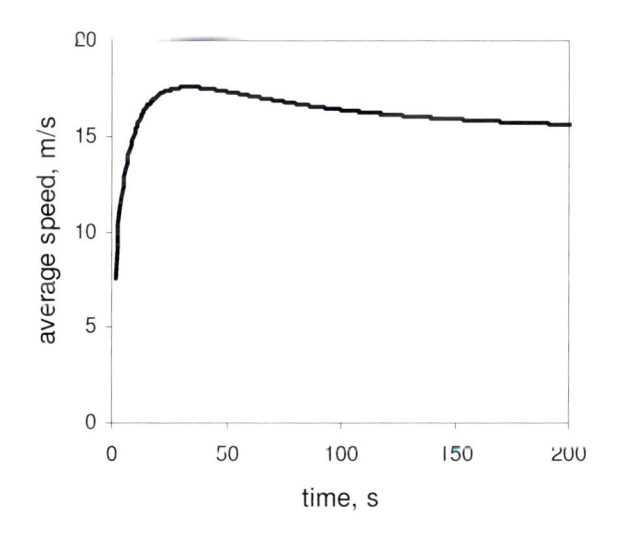

Figure 2.4 Predicted average speed as a function of race time

2.5 Further implications of the model

The times predicted by equation (19) and shown by the curve in Figure 2.1 can be thought of as the time which would be taken by a "standard" horse carrying the "standard" weight w_g to complete the distance D on good going at a hypothetical "standard" track with a conformation which is the average of all the 35 British racecourses included in the analysis.

This equation can then be used as a benchmark for performance, and also to investigate the effect of variations from standard conditions. The times calculated using equation (19) will be referred to as "model" times.

A. Effect of weight on time

Horserace ratings are normally expressed in pounds. Adding weight slows down a horse, but there are many opinions on how much extra weight is needed to slow a horse down by 1 second in a five-furlong race. Dickinson (1997) uses $\Delta w/\Delta T = 20$lb/s at 5f, but states that other handicappers use values ranging from 15lb/s to 24lb/s. 25lb/s is also often used.

If a horse completes a 5f race in 1 second less than the standard time, on standard going, and carrying standard weight, then its rating would be +25, with respect to some arbitrarily chosen reference level, if $\Delta w/\Delta T$ is taken to be 25 lb/s.

$\Delta w/\Delta T$ must decrease as race distance increases. If an extra 25lb causes an increase in time of 1s for a 5f race then for a 10f race the time increase will be at least 2s, giving $\Delta w/\Delta T \leq 12.5$lb/s (at 10f).

The predicted effect of weight on time can be found easily from equation (19). The simplest way is to increase the weight carried by 1lb and calculate the increase in time ΔT. This gives the results shown in Table 2.2.

D (f)	$\Delta w/\Delta T$ (lb/s)
5	29.94
6	23.91
7	19.89
8	17.03
10	13.26
12	10.87
14	9.21
16	7.99

Table 2.2 Effect of weight on time for different distances

The data from Table 2.2 is shown graphically in Figure 2.5. Also shown is a curve corresponding to 25lb/s at 5f, which simply decreases inversely with distance.

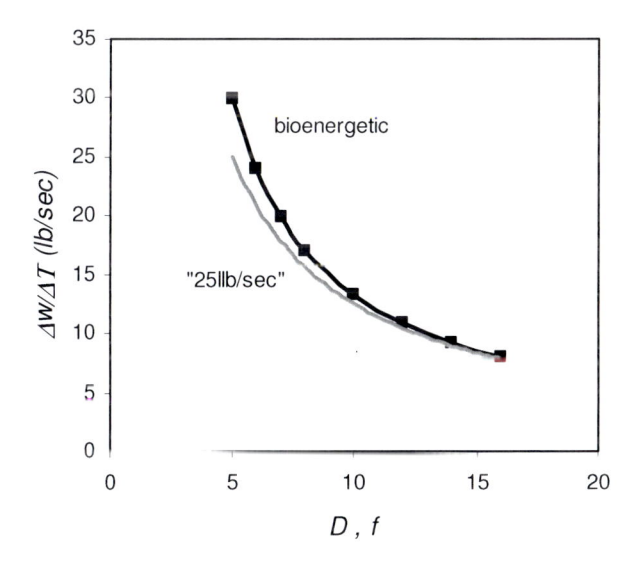

Figure 2.5 Effect of weight on time for different race distances (flat)

The model predicts $\Delta w/\Delta T$ somewhat higher than the "25lb/s at 5f" curve for the shorter distances, but at 16f the values are very close.

Although the predicted values of $\Delta w/\Delta T$ are of the same order of magnitude as those commonly used, they are somewhat higher, indicating that a little more weight is needed to slow a horse down than has been generally believed.

The model parameters could be changed to force lower $\Delta w/\Delta T$ values, but to get $\Delta w/\Delta T$ as low as 15lb/s at 5f would require values of $\dot{V}O_2 max$ much higher than have ever been reported for thoroughbreds.

B. Effect of weight on distance

The slowing-down effect of weight can also be expressed in terms of the change in distance covered caused by the extra weight, usually expressed in terms of pounds per length.

A length is the distance between the nose and the rear of a running horse, and is of the order of 8-9 feet (2.44-2.74m). In this work a length will be taken to be a constant 2.7m.

The British Horseracing Authority (BHA) states that "in Flat races one length is typically reckoned to be worth three pounds in sprint races, two pounds in mile races and one pound in staying races", while Dickinson (1997) gives 3-4 lb as the equivalent of one length for a 5f race.

If the relationship between weight and distance in lengths is written as $\Delta w/\Delta L$, then it can easily be derived from the values of $\Delta w/\Delta T$ given in the previous section. For each race distance, the average speed in m/s is calculated, and this is then converted to a speed in lengths per second, by dividing by 2.7. We then get

$$\frac{\Delta w}{\Delta L} = \frac{\Delta w/\Delta T}{\Delta L/\Delta T} \qquad (20)$$

The results are shown in Table 2.3 and Figure 2.6.

D (f)	T (s)	V (m/s)	$\Delta L/\Delta T$ (L/s)	$\Delta w/\Delta L$ (lb/L)
5	58.68	17.14	6.35	4.72
6	71.57	16.86	6.25	3.83
7	84.73	16.62	6.16	3.23
8	98.05	16.41	6.08	2.80
10	124.98	16.10	5.96	2.22
12	152.09	15.87	5.88	1.85
14	179.28	15.71	5.82	1.58
16	206.52	15.58	5.77	1.39

Table 2.3 Effect of weight on distance

The BHA define 5f, 6f and 7f as sprint races, and the average predicted $\Delta w/\Delta L$ for these three distances is 3.95 lb/L, which is in excellent agreement with values in common use, but for 5f races the value is significantly higher. However the average speeds for 6f and 7f are lower than 5f, so 6f and 7f races cannot be sprints in the sense of Keller.

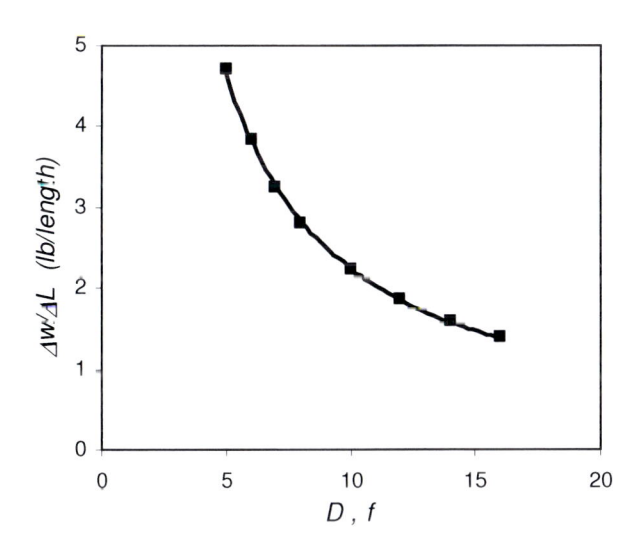

Figure 2.6 Pounds per length for different race distances (flat)

C. Effect of going

The values of $\Delta w/\Delta T$ and $\Delta w/\Delta L$ given above are for standard conditions. If the conditions are non-standard, the values change.

If the going is non-standard, the cost of running C_R changes, increasing as the going gets heavier and vice-versa. There is no effect on the air resistance or kinetic energy terms on the right-hand side of equation (19).

Changes in going can be represented by replacing C_R by $F_G C_R$ where F_G is the *going coefficient*. Experience shows that a range of $0.9 < F_G < 1.1$ covers as wide a variation as is likely to occur in practice.

F_G	\multicolumn{5}{c}{$\Delta w/\Delta T$ (lb/s)}
	0.9	0.95	1	1.05	1.1
	firmer	*firmer*	*standard*	*softer*	*softer*
5f	34.81	32.24	29.94	27.86	25.99
6f	27.88	25.78	23.91	22.24	20.74
7f	23.21	21.45	19.89	18.50	17.26
8f	19.87	18.36	17.03	15.85	14.80
10f	15.46	14.29	13.26	12.35	11.54
12f	12.66	11.71	10.87	10.13	9.47
14f	10.73	9.92	9.21	8.59	8.04
16f	9.31	8.61	7.99	7.46	6.98

Table 2.4 Effect of going on $\Delta w/\Delta T$

Table 2.4 gives values of $\Delta w/\Delta T$ obtained from equation (19) for different going coefficients, and these results are shown graphically in Figure 2.7. For heavy going, less weight is required to slow a horse down by one second, while for firm ground the opposite is true.

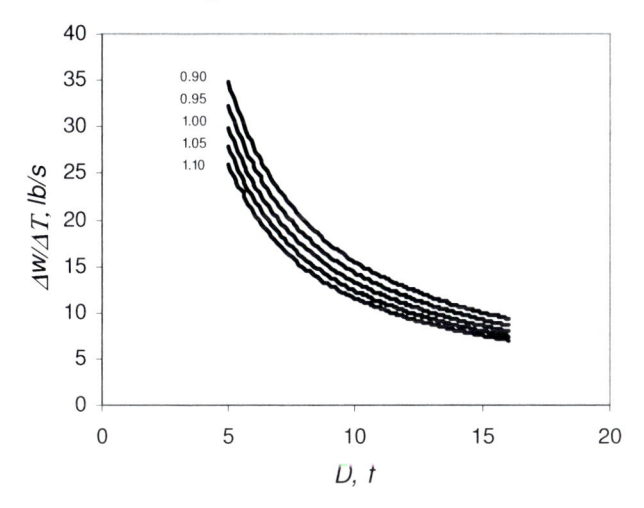

Figure 2.7 $\Delta w/\Delta T$ versus race distance for different going coefficients

D. Going allowances

When evaluating racehorse performance in terms of the time taken to complete the distance, it is necessary to allow for the effect of the going. This is done by comparing the times recorded at a given race meeting with standard times taking into account the ability of the horses running. The methods will be dealt with in more detail in Chapter 3.

There are two main methods in current use of allowing for the going.

The first is to work out a correction in terms of the number of *seconds per furlong (SPF)*, and modify the standard time to allow for this. For example if the standard time for a one-mile race at a particular course is 100.0s and the going slower than standard, on the soft side, with an allowance of 0.2 s/f, the standard is modified to 100.0 + (8 × 0.2) = 101.6s.

	going allowance SPF (s / furlong)				
F_G	0.9	0.95	1	1.05	1.1
	firmer	firmer	standard	softer	softer
5f	-0.681	-0.347	0	0.361	0.735
6f	-0.731	-0.372	0	0.386	0.786
7f	-0.768	-0.391	0	0.405	0.824
8f	-0.797	-0.406	0	0.420	0.852
10f	-0.838	-0.426	0	0.439	0.891
12f	-0.864	-0.439	0	0.451	0.915
14f	-0.882	-0.448	0	0.460	0.932
16f	-0.896	-0.454	0	0.466	0.944

Table 2.5 Variation in going allowance expressed in s/furlong

Using the bioenergetic model the change in time produced by different going coefficients, and the number of seconds per furlong required to adjust the standard time can be calculated. The results are shown in Table 2.5.

The key result here is that for non-standard going, represented by a fixed value of F_G, SPF is not constant, but increases significantly as the race distance increases. For a 2m race SPF needs to be about 30% higher than for 5f. The use of a fixed value of SPF for all distances results in a significant error in the correction required to the standard time.

A second method often used for allowing for the going is to add or subtract a constant weight correction Δw_G to the calculated rating when expressed in pounds.

Δw_G can be calculated using the model for standard conditions, by first finding the departure from standard time ΔT with a going factor F_G, and then multiplying by $\Delta w / \Delta T$. The results are shown in Table 2.6.

	Δw_G (lb)				
F_G	0.9	0.95	1	1.05	1.1
	firmer		standard	softer	
5f	-118.5	-56.0	0.0	50.2	95.5
6f	-122.2	-57.6	0.0	51.5	97.8
7f	-124.8	-58.8	0.0	52.5	99.6
8f	-126.8	-59.6	0.0	53.2	100.9
10f	-129.5	-60.8	0.0	54.2	102.8
12f	-131.3	-61.6	0.0	54.9	104.0
14f	-132.5	-62.2	0.0	55.3	104.9
16f	-133.4	-62.6	0.0	55.7	105.5

Table 2.6 Weight shift needed to allow for going

Again the correction required for a given F_G is not constant, but increases with race distance. However the variation is much less than that found for *SPF* (the 2m value of Δw_G is only 10% higher than 5f value), so if ratings are to be calculated in pounds this method of correcting for the going is preferable to the seconds per furlong method.

E. Effect of pace

Table 2.3 shows that the average speed for a standard 2mile (16f) race is only 9% slower than for a 5f race. Comparing Figure 2.4 with Figure 1.9 shows that the decrease in speed as distance increases is much less marked for horses than for humans. To re-quote Lloyd "horses ... keep it up much better". This means that the pace of a horse race needs to be judged to within a few per cent accuracy if a fast performance is to be achieved.

This is illustrated in Figure 2.8, in which the bend in the standard *D vs T* characteristic has been greatly exaggerated to make the graph clearer.

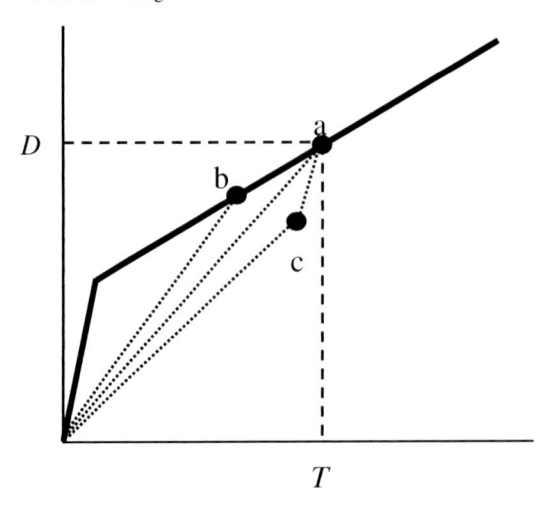

Figure 2.8 Effect of pace

Consider three standard horses (i.e. of equal ability), which are just capable of running a race in the standard time, which represents the limit of their ability.

Horse *a* runs at the correct pace, at the standard speed, and completes the distance D in the time T, when the point *a* is reached.

Horse *b* sets off at too fast a pace, and if this is maintained, reaches the exhaustion limit at point *b*, and then loses muscle power and will drop out of the reckoning.

Horse *c* sets off too slowly, and if it continues at this pace, will reach the point *c*. It will then need to accelerate to the speed of a sprinter (slope of the line *c-a*) in order to catch horse *a*. This is unlikely near the end of a race. If horse *c* continues beyond point *c* at the original pace, it will be physically impossible to catch horse *a*. Horse *c* has been left "too much to do".

In this example horse *a* has beaten two other horses of equal ability, simply by running at the right pace.

Inspection of the real D vs T characteristic shown in Figure 2.1 shows how little the slope of the curve changes for horses, which emphasises the critical importance of pace. A numerical example using equation (19) will illustrate this. If the race is a standard 1mile (8f) race and a standard horse b sets off at a speed only 0.5% higher than the standard speed, it will have used up all the available energy when a point 84m short of the finishing line is reached.

Of course the exhaustion limit in reality is a zone which is entered, rather than the sharp discontinuity implied by the model. However the sudden tailing off of a horse in the final stages of a race is a familiar sight. The correct judgement of pace for a given horse, course, and ground conditions is the most important skill a jockey can have.

The penalty for "going off too fast" is high, which explains why so many races are run at speeds well below the standard.

F. Getting the trip

It is common to hear racehorse trainers say that their horse didn't "get the trip" when it failed to perform well at a distance longer than one at which it had been previously successful. To the outsider, this phrase seems puzzling at first. Surely if a horse is capable of running over 8f, it will also be able to run 10f, albeit at a slightly slower pace?

The model parameters given in Table 2.1 define a "standard" horse, with fixed values of E_0 and P_m. However individual horses at peak fitness are characterised by their own values of these parameters. Horses with higher anaerobic stored energy E_0 tend to perform better at shorter distances while those with higher aerobic capacity P_m tend to be better at longer distances.

This is illustrated in Figure 2.9, which shows the limiting D vs T characteristics of two closely matched horses a and b. Horse a has a higher anaerobic capacity than horse b but a lower aerobic capacity, which results in a lower slope of the curve to the right-hand side of the bend. Again the bend in the D vs T characteristic has been exaggerated, for clarity.

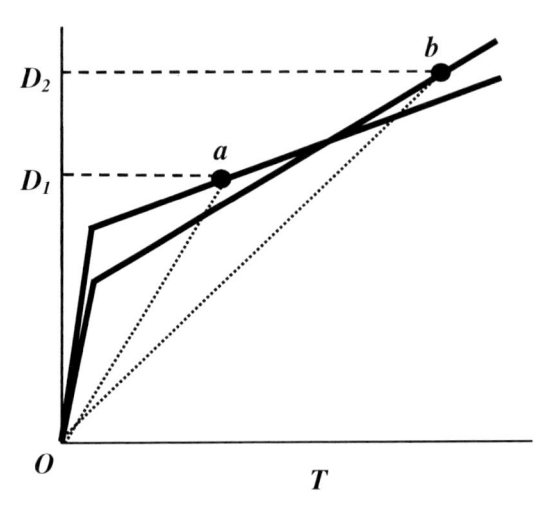

Figure 2.9 Effect of different slopes of D vs T characteristic

If for a race distance D_1 both runners set off at a speed represented by the slope of the line Oa horse a will win, because b will reach its exhaustion limit before distance D_1 has been reached. On the other hand, for a distance D_2 the reverse is true. Because the two distance/time characteristics cross each other, a will now reach its exhaustion limit before b.

This is only true for closely-matched horses. If one horse is so much better that the curves do not cross, it can win regardless of distance.

Referring again to the real D vs T characteristic shown in Figure 2.1, with its very small change in slope at the bend, it becomes clear that only very slight differences in E_0 and P_m are required, to determine which horse in a particular race will last the distance better than its competitors.

G. Gain in altitude

If the winning post is located at an altitude H metres higher than the starting gate, then an additional potential energy $(W+w)gH$ will have to be supplied, and needs to be added to the right-hand side of equation (17). For the power balance equation (19), the term to be added is $(1+w/W)gH/T$.

Checking height changes from maps gives H and the changes in times which result can be then calculated.

H. Other effects

$\Delta w/\Delta T$ also varies slightly as the weight carried w changes. A decrease of about 0.6 lb/s is produced as w is increased from 7st to 10st. (In all these analyses the weight of the horse W is taken to be constant. The effect of variations in W will be considered in Chapter 4.)

2.6   Jump racing

There are two main types of races run over jumps on turf in Britain, *hurdles* and *chases*, and they are usually run over distances between 2miles and 4miles.

In hurdles there are light timber obstacles at least 3' 6" high which are supposed to be cleared, but as in athletics, they can be knocked over without necessarily putting an end to a competitor's chances. Chases require the horse to clear solid plain fences at least 4' 6" high plus possibly a water jump, and so standard speeds for chases are slower than for hurdles.

The cost of running C_R is higher than for flat racing. The nature of the ground is different, and the jump racing season spans the winter months, during which the ground is generally heavier than during the summer months when flat racing is staged. Secondly, additional energy is needed to clear the obstacles. Both of these effects can be represented by an increase in the value of C_R. Since the number of obstacles jumped increases with distance D (there are about 4 fences per mile for hurdles and 6 fences per mile for chases), they can be approximated by an energy cost which increases with D.

The calculation procedures are the same as for flat racing, but for jump racing, the weight of the horse has been assumed to be 1200lb, as these horses are older and generally heavier (see Chapter 4), while the standard weight has been taken to be 12st = 168lb.

All other parameters for the model are assumed to be the same as given in Table 2.1, except for C_R, which has to be determined separately for hurdles and chases.

A. Hurdles

For determination of the single unknown parameter C_R the data set consisted of 167 standard times for distances from 16f to 28f (2miles to 3.5miles).

Using the stochastic optimisation method C_R = 4.0053 gave the best fit to the data (standard error = 0.0199). Figure 2.10 shows the calculated distance-time curve, with the cloud of data points in the background.

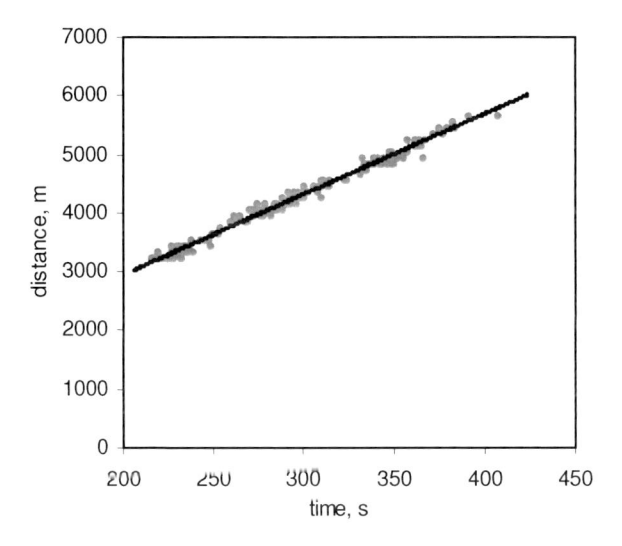

Figure 2.10 Distance versus race time (hurdles)

Table 2.7 shows the effect of weight carried on time and lengths for key distances common in hurdle racing.

D (f)	T (s)	V (m/s)	$\Delta L/\Delta T$ (L/s)	$\Delta w/\Delta T$ (lb/s)	$\Delta w/\Delta L$ (lb/L)
16	221.59	14.53	5.38	7.80	1.45
20	280.22	14.36	5.32	6.18	1.16
24	338.91	14.25	5.28	5.12	0.970
28	397.64	14.16	5.25	4.36	0.832

Table 2.7 Effect of weight on time for different distances (hurdles)

B. Chases

For chases the data set comprised 221 standard times, for distances from 16f to 36f (2miles to 4.5miles). In this case the optimum value of C_R was 4.2211 (standard error = 0.0165). The calculated distance-time characteristic is shown in Figure 2.11. The data point on the extreme right-hand side of the graph is the Grand National, for which the standard time is longer than the average prediction of the model. This is due to the fact that the Grand National fences are more challenging than average steeplechase fences.

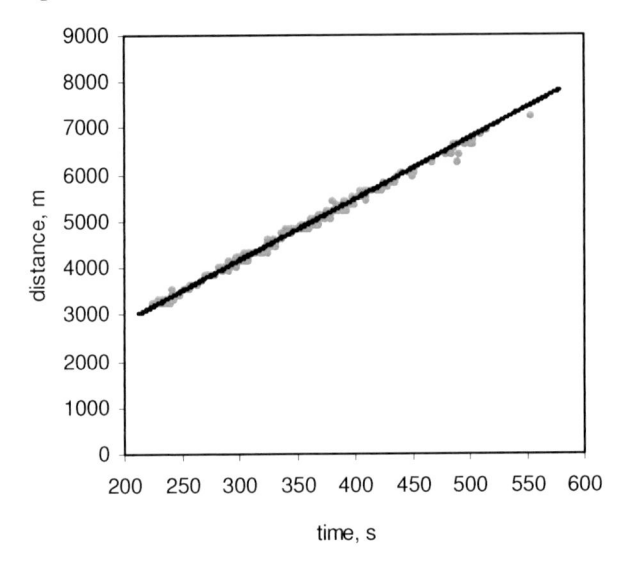

Figure 2.11 Distance versus race time (chases)

Table 2.8 shows the effect of weight carried on time and lengths for key distances in chasing.

D (f)	T (s)	V (m/s)	$\Delta L/\Delta T$ (L/s)	$\Delta w/\Delta T$ (lb/s)	$\Delta w/\Delta L$ (lb/L)
16	230.85	13.94	5.16	7.25	1.40
20	291.99	13.78	5.10	5.75	1.13
24	353.18	13.67	5.06	4.76	0.940
28	414.41	13.59	5.03	4.06	0.807
32	475.66	13.53	5.01	3.54	0.707
36	536.92	13.49	5.00	3.14	0.629

Table 2.8 Effect of weight on time for different distances (chases)

The results show that for jump racing one second corresponds to 7.8 to 3.1lb according to distance.

Figure 2.12 illustrates the variation of $\Delta w/\Delta L$ with race distance for hurdles (upper points) and chases (lower points). The values for chases are slightly lower. It is often stated that one pound corresponds to one length for jump racing, and this is seen to be roughly true. However, $\Delta w/\Delta L$ falls from 1.45 lb/L at 2miles to less than half this value (0.63 lb/L) at the Grand National distance.

The values of $\Delta w/\Delta T$ and of $\Delta w/\Delta L$ given in Table 2.7 and Table 2.8 and shown in the graphs are for standard going only. As is the case for flat racing they will be higher if the going is firmer and lower if the going is softer.

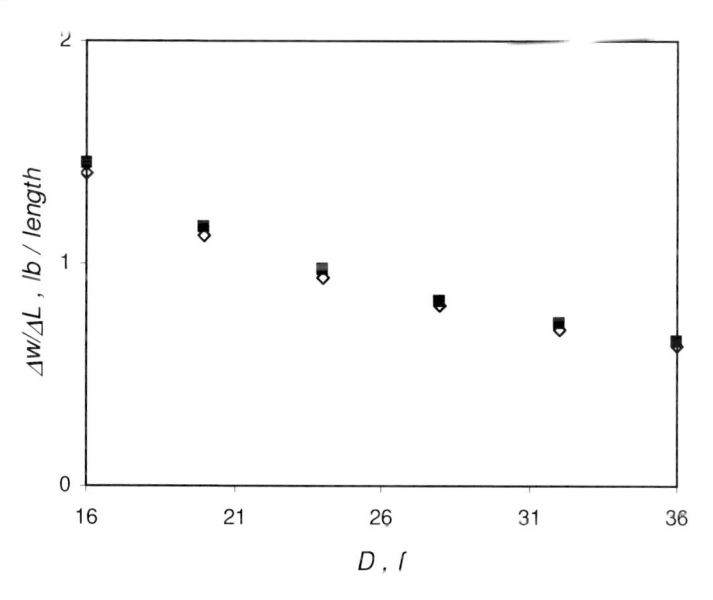

Figure 2.12 Pounds per length as a function of jump race distance

For jump racing the energy supply is dominated by aerobic energy, which increases from about 83% of the total at 16f to about 91% of the total for the longest distance races.

Energy usage is dominated by the locomotion energy (cost of running)

which is about 87% to 89% of the total. Air resistance decreases very slightly with distance, accounting for about 9% of the total, while the kinetic energy term is very small, at 3% to 1.5%.

2.7 Conclusion

In this chapter the Peronnet and Thibault model has been extended to the running of racehorses, including the effect of variation of weight carried. The model parameters have been evaluated for flat racing, hurdles, and chases by fitting to standard times for race distances run on turf in Britain. The model allows the effect of weight carried on race time and distance in lengths to be calculated, and also the effect of the going.

In this work standard times from the *Racing Post* have been used, but other sets of standard times have also been tested. This produced small changes in the calculated model parameters and calculated dependencies, but no substantial change to the conclusions reached.

Chapter 3. Assessment of Performance - Time-Based Ratings

The model developed in Chapter 2 gives the time which would be recorded by a "standard horse" when running at a "standard track". This standard horse always puts in a performance governed by the model, whatever the race distance, and is characterised by the two parameters E_0 and P_m. Such a horse does not (and could not) exist, but it serves as the benchmark against which the performance of real horses can be judged. (T_E and T_P are considered to be constants).

If a horse completes a race over a specific track and distance in a given time T we cannot calculate the values of both E_0 and P_m from this. To do so requires several sets of times *at different distances* and fitting the data to the model in the same way as was done to determine the parameters for the standard horse.

Determination of E_0 and P_m has been done (Peronnet and Thibault 1989) for certain elite individual athletes, based on their personal best performances at several different distances. However there is no possibility of gathering such data for each of the thousands of racehorses in training.

The solution is to adopt the method used by Keller (1973), who suggested that the relationship between E_0 and P_m is fixed. A trained athlete with a high P_m will have a correspondingly high value of E_0, and vice versa. For the standard horse $E_0/P_m = T_0 = 2615.3/68.296 = 38.29$s, considerably lower than the value for male athletes (see section 1.4).

Use of a constant T_0 means that a specific performance can be represented by a single performance measure, either E_0 or P_m. The choice is arbitrary, but as the aerobic pathway is the dominant source of energy in horse racing it is preferable to use P_m.

Putting $E_0 = P_m T_0$ in equation (19), we get

$$P_m\left(\frac{T_0}{T}(1-e^{-T/T_E}) + 1 - \frac{T_P}{T}(1-e^{-T/T_P})\right) = \left(1+\frac{w}{W}\right)C_R\frac{D}{T} + k\left(\frac{D}{T}\right)^3 + 2\left(1+\frac{w}{W}\right)\left(\frac{D}{T}\right)^2\frac{1}{T}$$

... (21)

For a race run over a distance D in a time T on a track for which the cost of running is C_R' and with a going coefficient F_G, the average power P is obtained from (21), as

$$P = \frac{1}{F(T)}\left[\left(1+\frac{w}{W}\right)F_G C_R'\frac{D}{T} + k\left(\frac{D}{T}\right)^3 + 2\left(1+\frac{w}{W}\right)\left(\frac{D}{T}\right)^2\frac{1}{T}\right]$$

... (22)

which gives the measure of performance as a **power rating,** in units of Wkg^{-1}. This represents an *apparent* metabolic power with a corresponding *apparent* $\dot{V}O_2 max$. Equation (22) will be frequently referred to both for the calculation of ratings and their usage, so for convenience it will be referred to in future as *the Power Equation*.

$\dot{V}O_2 max$ as measured on a treadmill is not generally a good indicator of running performance (Rose et al, 1990). A horse with a lower $\dot{V}O_2 max$ may have a superior cardiovascular system, muscle efficiency, and running gait when compared with a horse with a higher $\dot{V}O_2 max$ but which has inferior systems for translating this into speed across the ground. Calculation of P using the Power Equation assumes that all these energy conversion pathways are lumped together in series, and the result represented by a single measure. Thus the value of P calculated using the Power Equation, and its associated $\dot{V}O_2 max$ should be thought of as 'apparent' metabolic values, calculated from the measurable quantities D and T, and which include all these energy conversion pathways.

The dimensionless quantity $F(T)$ is a function of the race time T and the constants T_0, T_E and T_P, and is given by

$$F(T) = \frac{T_0}{T}(1 - e^{-T/T_E}) + 1 - \frac{T_P}{T}(1 - e^{-T/T_P}) \qquad (23)$$

$F(T)$ decreases from about 1.4 to 1.04 over the range of race times found in horse racing (see Figure 3.1). It is related to the changing balance between the anaerobic and aerobic power as time increases.

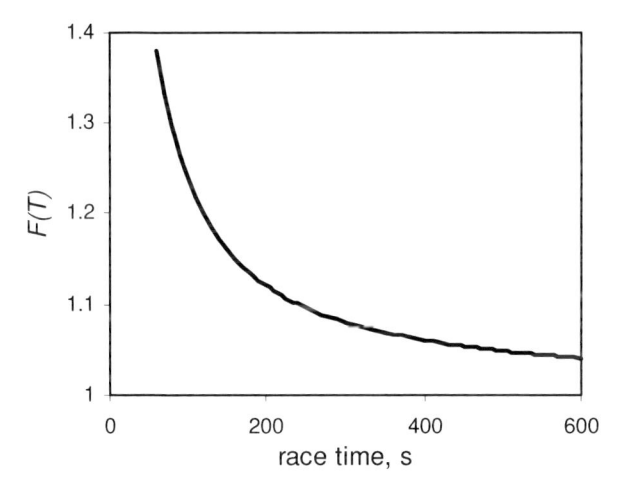

Figure 3.1 Variation of $F(T)$ with race time

3.1 Cost of running at a specific track C_R'

If the standard time at a specific track (T_S) is greater than the model time, the cost of running C_R' will be higher than the standard value, and vice-versa. The required value is obtained easily, by solving the Power Equation for C_R, with $T = T_S$, $w = w_s$, $F_G = 1$ and $P = P_m$, which gives

$$C_R' = \frac{P_m F(T_s) - k\left(\frac{D}{T_s}\right)^3 - 2\left(1+\frac{w_s}{W}\right)\left(\frac{D}{T_s}\right)^2 \cdot \frac{1}{T_s}}{\left(1+\frac{w_s}{W}\right)\frac{D}{T_s}} \quad (24)$$

3.2 Beaten horses

Calculation of the power rating P requires knowledge of the time T, which is published in Britain for the race winner, but unfortunately not for the beaten horses, even though these times are available. Instead a set of "distances beaten" is given. The distances are selected from a fixed set, 1 length, 1¼ lengths, 1½ lengths, 1¾ lengths, 2 lengths, 2½ lengths etc. Distances less than 1 length are defined as a *nose*=0.02 lengths, a *short head* = 0.05 lengths, a *head* = 0.1 lengths, and a *neck* = 0.25 lengths.

Since 2008 these distances are calculated from the Photo Finish computer on the assumption that the finishing speeds on turf tracks are as shown in Table 3.1.

Race type	Official Going	Finishing speed, V_F (lengths per second)
Flat	Good or faster	6.0
Flat	Good(good to soft in places)	5.5
Flat	Soft or softer	5.0
Jumps	Good or better	5.0
Jumps	Good to soft or softer	4.0

Table 3.1 Official finishing speeds

So if the winner's time is T_W, the time for a beaten horse can be calculated from the published data, as

$$T = T_W + \frac{d}{V_F} \quad (25)$$

where *d* is the total (cumulative) distance beaten. Since the values of *d* are chosen from a set of discrete values there is an unnecessary loss of precision in the calculation. In this system a length is not a fixed physical distance, but varies from race to race.

3.3 Time-based ratings

The determination of F_G is the key to calculation of time-based power ratings using the Power Equation.

A recent development in British racing has been the arrival of the GoingStick, a device which is pushed into the ground to measure penetration and then moved sideways to measure shear. The results are output on a scale from 1 to 15. Low numbers (<7) are obtained for ground on the softer side of good, while high numbers (>8) correspond to firmer ground. From the start of 2009 all courses must publish average GoingStick readings based on at least 3 readings for each furlong of the course.

GoingStick readings are very useful for trainers, owners and punters, who now have an objective measurement of the state of the ground, which is preferable to the old system of classification as "Good", "Soft" and so on, which were subjective judgements made by the Clerks of the Course, and which were often the subject of dispute.

Initial experience shows that while GoingStick readings at a given course give consistent results, it is difficult to compare values at different courses, due to differences in the type of turf, soil, and permeability. At present there is no way to convert GoingStick readings to a going allowance which is sufficiently accurate for the calculation of time-based ratings, so allowance for the going must be made by other methods.

Consider a race which is run competitively, but is won in a time significantly slower than standard. This could be due to the fact that (a) the pace of the race was slow (b) the going was softer than standard, or (c) the horses running were of lower ability than standard. It is necessary to separate and quantify these three effects.

This can be done by considering all the races run on a particular day at a meeting, looking at the best performance(s) on that day, and comparing with previous best performances. The general procedure is as follows :

1. Calculate C_R' for each race distance at the meeting, using equation (24)

2. Calculate a provisional power rating (P) for each horse using the Power Equation - with $F_G = 1$

3. Look up the previous best rating P^* achieved by each horse.

4. For each horse, calculate the going coefficient F_G which would be required to adjust P to agree with P^* (i.e. for each horse to have run to its previous best)

5. Select a sample group of the fastest horses in each race. Typically, this should be the winner plus horses which finish within (say) 10 lengths of the winner (flat) or 20 lengths of the winner (jumps). The sample size should not exceed six - horses not finishing in the first six will probably not have run close to their previous best.

6. Calculate the average value of F_G for the sample group in each race.

7. Select the lowest of the set of average values of F_G

8. Apply this value of F_G to all races, re-calculating all power ratings

The slower the race, the greater will be the value of F_G required for that race to bring the fastest horses ratings up to their previous best. If F_G for the slower races were to be taken as representative of the meeting as a whole, then the calculated ratings for horses in the other races would be far in excess of what they had ever achieved previously. For this reason the lowest average value of F_G is used. However careful subjective judgement is needed for each meeting to check that the best set of samples has been used.

The key to the above method is **step 4**. The provisional power rating P is given by the Power Equation with $F_G = 1$. If F_G is adjusted to give the target value P^*,

$$P^* = \frac{1}{F(T)}\left[\left(1+\frac{w}{W}\right)F_G C_R' \frac{D}{T} + k\left(\frac{D}{T}\right)^3 + 2\left(1+\frac{w}{W}\right)\left(\frac{D}{T}\right)^2 \frac{1}{T}\right]$$

$$\ldots (26)$$

subtracting (22) from (26) gives

$$P^* - P = \Delta P = (F_G - 1)\frac{1}{F(T)}\left(1+\frac{w}{W}\right)C_R' \frac{D}{T}$$

from which the required going coefficient is obtained, as

$$F_G = 1 + \frac{\Delta P\, F(T)}{\left(1+\frac{w}{W}\right)C_R' \frac{D}{T}} \qquad (27)$$

Bioenergetics and Racehorse Ratings   Bob Wilkins

3.4 Example

The method is illustrated by the following example, representing a fictitious flat race meeting with 6 races on the card. (Note - these results are however derived from real flat races. The names of the horses and some of the data have been altered to give a set of races which illustrates the main points to be made in connection with the calculation of ratings).

Race 1. 2yo stakes, 7 runners, 5f (D =1005.8m), T_S = 57.8s, C_R' = 3.6347 (a)

P^* (b)	horse	btn (c)	d (d)	w (e)	T (f)	P (g)	ΔP (h)	F_G (i)	P'
-	Felix	0	0	129	62.35	62.25	-	-	64.07
65.20	Secrets	3½	3.5	129	62.99	61.48	3.72	1.0775	63.29
65.10	Lancashire Lass	1¾	5.25	124	63.30	60.89	4.23	1.0886	62.67
64.71	Fee Bee	shd	5.3	124	63.31	60.87	3.83	1.0806	62.66
63.97	Flautist	1½	6.8	129	63.59	60.78	3.19	1.0671	62.57
63.53	Cricket Man	½	7.3	129	63.68	60.67	2.86	1.0601	62.46
-	Susie Wong	2	9.3	124	64.04	60.04	-	-	61.81
							ave	1.0748	

Notes:

(a) Given the standard time, the cost of running this distance at this track is given by equation (24)

(b) P^* is the previous best power rating this horse has achieved

(c) btn gives the distances beaten in conventional format (between successive horses)

(d) d is the cumulative distance beaten, i.e. behind the winner

(e) w is the weight carried (129 = 9st 3lb etc)

(f) T is the time taken. For the winner this is known, for beaten horses it is calculated from Table 3.1 and equation (25), assuming an official going of Good to Soft

(g) P is the provisional power rating, calculated with $F_G = 1$

(h) ΔP is the difference between P^* and P

(i) F_G is the going coefficient required to bring each horse's power rating up to its previous best, calculated using equation (27).

1.0748 is the average of the going coefficients needed for horses in the sample group. (These are shown shaded in each race)

Race 2. 4yo+ handicap, 10 runners, 9f (D=1810.44m), T_S = 110.5s, C_R' = 3.6796

P*	horse	btn	d	w	T	P	ΔP	F_G	P'
67.66	Glenniesroth	0	0	123	114.23	65.50	2.16	1.0401	67.52
67.80	Dundee Man	4½	4.5	130	115.05	65.27	2.53	1.0470	67.30
67.61	Oscars List	nse	4.52	134	115.05	65.46	2.15	1.0398	67.49
67.07	Needless	1½	6.02	132	115.32	65.18	1.89	1.0350	67.21
68.20	Oui Oui	½	6.52	125	115.42	64.79	3.41	1.0636	66.81
67.66	Penguin Blue	nk	6.77	131	115.46	65.04	2.61	1.0486	67.07
68.89	Sailboat Ride	1¾	8.52	125	115.78	64.55	4.33	1.0812	66.56
66.67	Moon Stream	2	10.52	117	116.14	63.94	2.73	1.0517	65.93
68.15	Seeking	½	11.02	129	116.23	64.44	3.71	1.0695	66.45
68.00	Rhythm Stick	1¼	12.17	134	116.44	64.54	3.46	1.0647	66.55

                                                                         ave **1.0457**

Race 3. 4yo+ handicap, 9 runners, 16f (*D*=3218.56m), T_S = 210s, C_R' = 3.8155

P*	horse	btn	d	w	T	P	ΔP	F_G	P'
68.15	Whisky Chaser	0	0	134	225.60	63.18	4.97	1.0897	65.26
67.46	Birdman	¾	0.75	122	225.74	62.58	4.88	1.0890	64.64
67.61	Old Field	3	3.75	124	226.28	62.50	5.11	1.0933	64.56
68.20	Cheshire Cat	1½	5.25	136	226.55	62.97	5.23	1.0947	65.04
67.12	Zonguldak	2¾	8	118	227.05	61.98	5.14	1.0946	64.02
67.22	Three Dances	nk	8.25	122	227.10	62.15	5.07	1.0930	64.20
67.95	My Spirit	2¼	10.5	136	227.51	62.66	5.29	1.0961	64.73
67.51	Welsh Rock	2¼	12.75	127	227.92	62.12	5.39	1.0988	64.17
67.07	Concordial	32	44.75	116	233.74	59.86	7.21	1.1365	61.84
							ave	**1.0924**	

Race 4. 2yo seller, 10 runners, 6f (*D*=1206.96m), T_S = 74.22s, C_R' = 3.6481

P*	horse	btn	d	w	T	P	ΔP	F_G	P'
64.91	Marbella	0	0	118	74.22	63.86	1.04	1.0207	65.75
64.56	Open Sesame	8	8	118	75.67	62.38	2.18	1.0440	64.25
64.76	Glastonbury	½	8.5	123	75.77	62.52	2.24	1.0450	64.39
64.41	Alligator Boy	2¼	10.75	118	76.17	61.89	2.52	1.0512	63.74
63.78	Vermont Holiday	½	11.25	111	76.27	61.49	2.29	1.0467	63.33
64.46	Two Of Us	1¼	12.5	118	76.49	61.58	2.88	1.0586	63.43
65.00	Declaration	1	13.5	123	76.67	61.63	3.37	1.0685	63.48
64.36	Chessman	3¼	16.75	123	77.27	61.07	3.29	1.0673	62.91
63.87	Connelly Street	8	24.75	123	78.72	59.74	4.14	1.0858	61.55
-	Catkins	dht	24.75	118	78.72	59.52	-	-	61.32
							ave	**1.0366**	

Race 5. 3yo+ handicap, 13 runners, 5f (*D*=1005.8m), T_S = 57.8s, C_R' = 3.6347

P*	horse	btn	d	w	T	P	ΔP	F_G	P'
67.95	Victorious Nag	0	0	133	61.08	64.03	3.92	1.0795	65.89
67.95	Everest	1½	1.5	135	61.35	63.77	4.18	1.0850	65.62
68.20	Harlow Carr	shd	1.55	130	61.36	63.53	4.67	1.0953	65.37
68.49	Boardwalk	hd	1.65	133	61.38	63.65	4.85	1.0987	65.49
67.80	Steel Cutter	nk	1.9	133	61.43	63.59	4.22	1.0860	65.43
68.89	Himalayan Path	nse	1.92	146	61.43	64.18	4.71	1.0950	66.04
67.12	Narhdo	1¼	3.17	135	61.66	63.39	3.73	1.0762	65.23
67.80	Sun and Sand	1½	4.67	131	61.93	62.86	4.94	1.1016	64.69
68.30	Naajad	½	5.17	136	62.02	62.97	5.32	1.1091	64.81
67.66	Master Reynard	½	5.67	133	62.11	62.73	4.93	1.1014	64.55
68.00	Scimitar	½	6.17	121	62.20	62.07	5.93	1.1233	63.88
67.76	Bullfighter	2¾	8.92	133	62.70	62.00	5.75	1.1192	63.82
67.95	Spirit Level	½	9.42	130	62.79	61.76	6.19	1.1288	63.57

ave **1.0899**

Race 6. 4yo+ handicap, 8 runners, 8f (*D*=2413.92m), T_S = 152s, C_R' = 3.7248

P*	horse	btn	d	w	T	P	ΔP	F_G	P'
66.87	Chris D	0	0	119	158.20	64.81	2.07	1.0376	66.87
66.97	Bill Bones	7	7	117	159.47	64.10	2.87	1.0526	66.15
67.90	Maalesef	1¼	8.25	129	159.70	64.56	3.34	1.0609	66.62
67.46	Orange Canyon	2¾	11	127	160.20	64.23	3.23	1.0591	66.29
67.41	Red Sox Fan	2¾	13.75	122	160.70	63.76	3.65	1.0672	65.81
67.51	Golden Wolf	nk	14	132	160.75	64.21	3.00	1.0003	66.27
67.02	Woodside Ferry	3	17	114	161.29	63.12	3.90	1.0725	65.14
66.67	Baby Basket	1	18	120	161.47	63.31	3.36	1.0623	65.34

ave **1.0504**

final F_G **1.0376**

Glenmesroth put up the best performance of the day, with a raw power rating 65.50, 2.16 below its previous best - a going coefficient of 1.0401 would be needed to bring it up to its previous best. Using the criteria given previously, the lowest of the set of average going coefficients is found to be 1.0366 in race 4. However this was a two-year-old race in the early part of the season, and it is likely that the winner improved significantly on its previous best rating, which was low. In general, competitive handicaps usually provide the key to final selection of the going coefficient for the meeting, so a good solution is to use 1.0376, which will bring the rating of the winner of the last race (another handicap) up to match its previous best. The final adjusted power ratings P' in all the right-hand columns were calculated using the Power Equation with $F_G = 1.0376$.

Subjective judgement is needed in choosing the final value of F_G to be used at a meeting, but with experience this is usually not difficult. If the initial estimates in the previous example had been based on all horses finishing within 5 lengths of the winner, rather than 10 lengths, the same conclusion would have been reached. Often a small group of runners finish close together with the rest of the field well behind. Such a group is self-selecting as a sample.

Some time-handicappers calculate the going allowance based on the winner's time only. For our example meeting the data would be:

race	P*	horse	W	T	P	ΔP	F_G	P'
1	-	Felix	129	62.35	62.25	-	-	64.07
2	67.92	Glenniesroth	123	114.23	65.50	2.16	1.0401	67.52
3	68.44	Whisky Chaser	134	225.60	63.18	4.97	1.0897	65.26
4	65.03	Marbella	118	74.22	63.86	1.04	1.0207	65.75
5	68.23	Victorious Nag	133	61.08	64.03	3.92	1.0795	65.89
6	67.09	Chris D	119	158.20	64.81	2.07	1.0376	66.87

The lowest value of F_G is 1.0207 for Marbella, winner of race 4. However as mentioned previously, this is a two-year-old with a low previous best, and if this value were applied to all the runners the results would be very conservative. The winners of the handicap races 2 and 6 would be allotted ratings well below their previous best, despite winning in competitive company.

Again the solution is to use the next-lowest value of F_G, i.e. 1.0376 from race 6. This gives Marbella a significant increase in its best rating, but this is justifiable.

This example illustrates both the advantages and disadvantages of time-based ratings. Phil Bull, the founder of *Timeform* once remarked that time ratings do not tell you how good a horse is, they tell you "how bad he isn't". The winners of races 1 and 3 achieved low power ratings, possibly because the races were run at a slow pace, or possibly because their performance on the day was poor, so the information gained about their ability is not conclusive. On the other hand it is clear that the winner of race 4 showed a substantial improvement on previous performances.

The method requires storage and recall of the power ratings of all horses, so that the values of the best previous ratings P^* are available. Dickinson (1997) has described how to get a handicapping system started from scratch. Alternatively a set of ratings in pounds from another handicap system can be converted to power ratings using the method described in the next section, and then be used as a starting point.

Determination of power ratings needs high-precision computations (see Appendix B), but the results shown in the example only display 3-5 significant digits, for ease of reading. Furthermore, the power ratings are spread over a relatively narrow range deviating from the standard value $P_m = 68.296$. There are many ways of converting power ratings to a more readable form, with a wider spread. For example they can be expressed as percentage deviations from the standard, i.e. $100(P'-P_m)/P_m$. However a more familiar display output can be obtained if desired by converting them to a scale expressed in pounds, using the method described in the next section.

3.5 Conversion of power ratings to a scale given in pounds

Consider a horse which runs a race of distance D in a time T at a track with a cost of running C_R' and a going coefficient F_G. Its power rating will be given by the Power Equation. Suppose now that a standard horse runs in the same race, and dead-heats with our runner (i.e. finishes in the same time). Since the standard horse always generates a power rating P_m, it would have to be carrying a different weight, say w_m, where

$$P_m = \frac{1}{F(T)}\left[\left(1+\frac{w_m}{W}\right)F_G\,C_R'\,\frac{D}{T} + k\left(\frac{D}{T}\right)^3 + 2\left(1+\frac{w_m}{W}\right)\left(\frac{D}{T}\right)^2\cdot\frac{1}{T}\right]$$

... (28)

The difference between the weights carried $\Delta w = (w_m - w)$ is a measure of the difference between the two horses, expressed in pounds. If Δw is positive the horse is below standard, while Δw is negative the horse is above standard.

Subtracting the Power Equation from equation (28) we get

$$\Delta P = P_m - P = \frac{1}{F(T)} \left[\frac{(w_m - w)}{W} F_G C_R' \frac{D}{T} + 2 \frac{(w_m - w)}{W} \left(\frac{D}{T}\right)^2 \cdot \frac{1}{T} \right]$$

Note that the air resistance term disappears, giving

$$\Delta P = P_m - P = \frac{\Delta w}{F(T)W} \left[F_G C_R' \frac{D}{T} + 2 \left(\frac{D}{T}\right)^2 \cdot \frac{1}{T} \right]$$

so
$$\Delta w = q (P_m - P) \quad (29)$$

where

$$q = \frac{F(T) W}{F_G C_R' \frac{D}{T} + 2 \left(\frac{D}{T}\right)^2 \cdot \frac{1}{T}} \quad (30)$$

	q		
D (f)	$F_G=0.9$	$F_G=1$	$F_G=1.1$
5	22.382	20.447	18.820
6	22.422	20.431	18.765
7	22.436	20.406	18.713
8	22.436	20.378	18.665
10	22.428	20.330	18.591
12	22.417	20.294	18.538
14	22.406	20.266	18.499
16	22.396	20.244	18.469
average	22.415	20.349	18.633

Table 3.2 Factor for conversion from ΔP to Δw

q is almost constant for races run in the standard time, for distances from 5f to 16f on the flat, for a given value of F_G. This is shown in Table 3.2.

For a standard race on standard going q varies less than 1% from an average value of 20.349, so under these conditions, equation (29) gives a simple method for converting power ratings to pounds. However this is only to display the ratings in a more familiar form.

Δw is the difference between the actual performance and a standard reference value w_{REF}, which can be arbitrarily chosen, so that

$$\text{Rating in lb} = w_{REF} - \Delta w \qquad (31)$$

When time ratings are expressed in pounds, w_{REF} is often chosen to be 100. Alternatives are to use w_S (126lb for flat racing and 168lb for jump racing), or even a value of zero.

With w_{REF} chosen to be 100, the ratings of the winners in the previous example are as given in Table 3.3.

race	P*	horse	w	T	P'	q	Δw	rating (lb)
1	-	Felix	129	62.35	64.07	21.55	91.112	9
2	67.92	Glenniesroth	123	114.23	67.52	20.38	15.750	84
3	68.44	Whisky Chaser	134	225.60	65.26	20.82	63.103	37
4	05.00	Marbella	118	74.??	65.75	20.90	53.135	47
5	68.23	Victorious Nag	133	61.08	65.89	21.10	50.849	49
6	67.09	Chris D	119	158.20	66.87	20.37	29.035	71

Table 3.3 Winners ratings in pounds

The values of q and Δw were computed using $F_G = 1.0376$ and this results in the "speed" or "time" ratings, in the final column, rounded to the nearest pound.

It should be emphasised again that conversion to a pounds scale is done only to make the results more readable.

If it is required to convert an existing set of ratings in pounds to equivalent power ratings in W/kg, this can be done by solving (29) for

P and using (31) to obtain

$$P = P_m - \frac{w_{REF} - \text{Rating in lb}}{q} \qquad (32)$$

When doing this the appropriate values of F_G and C_R' will generally not be known, so q will have to be estimated using $F_G = 1$ and $C_R' = C_R$.

3.6 Weight-for-age scales

One of the objectives of Admiral Rous's experiments was to determine the weight allowance which should be given to young, immature horses, so that they could have a fair chance when competing against older horses. This resulted in a set of "weight-for-age" (WFA) tables, which have been modified over the years, but are still published officially by the BHA. For example, when racing over 5f in late July, a two year old is regarded as being 28lb below maturity. Handicappers often add this value to their calculated rating, so that horses of different ages can be compared directly.

Opinions differ as to whether addition of a WFA allowance to time-based ratings is necessary or desirable. On balance it seems that time-based ratings are a measure of what a horse has actually achieved and are better left as such.

Further consideration will be given to WFA allowances in Chapter 4.

Chapter 4. Assessment of Performance - Form-Based Ratings

The most common way of calculating racehorse ratings is to assess the performance of a horse *relative to the other runners in the same race* rather than by direct comparison with the clock. This produces ratings based on "collateral form" and enables much more information to be obtained from race results. In the meeting given as an example in the previous chapter, the winner of race 3, *Whisky Chaser* had a high previous best power rating, and beat a relatively high-class field, but was awarded a relative poor time-based rating, because the speed of the race was relatively slow - but for two-mile flat races like this, it is common for the pace to be slow, as discussed previously.

Power ratings based on collateral form rather than time can be obtained using a similar procedure to that described in Chapter 3. The difference is that the factor F_G is calculated for each individual race, rather than for the meeting as a whole. This is equivalent to saying that when a race is run in a relatively slow time, it could be due to the going, or due to a slow pace, and that the two causes are indistinguishable, as far as the calculation is concerned.

4.1 Example

Calculations of collateral form ratings for the races given in the example meeting are shown below. For each race a group of runners is selected as a sample, using the same guidelines for sample selection as for time-based ratings. The average value of F_G for the group is taken and then applied to all runners in that race. The runners finally selected for the sample group are shown shaded.

Race 1. 2yo stakes, 7 runners, 5f (D =1005.8m), $57T_S$=.8s, C_R' = 3.6347

$P*$	horse	btn	d	w	T	P	ΔP	F_G	P'
-	Felix	0	0	129	62.35	62.25	-	-	65.87
65.20	Secrets	3½	3.5	129	62.99	61.48	3.72	1.0775	65.07
65.10	Lancashire Lass	1¾	5.25	124	63.30	60.89	4.22	1.0886	64.44
64.71	Fee Bee	shd	5.3	124	63.31	60.87	3.83	1.0806	64.43
63.97	Flautist	1½	6.8	129	63.59	60.78	3.19	1.0671	64.34
63.53	Cricket Man	½	7.3	129	63.68	60.67	2.86	1.0601	64.23
-	Susie Wong	2	9.3	124	64.04	60.04	-	-	63.56

ave   1.0748

In race 1 the winner has no previous form, but within ten lengths there is a group of five for which the average F_G is 1.0748. After applying this to all runners, some of the sample group are awarded ratings above their previous best, while others are below, and a clear rating is obtained for the winner.

Race 2. 4yo+ handicap, 10 runners, 9f (D=1810.44m), T_S = 110.5s, C_R' = 3.6796

P*	horse	btn	d	w	T	P	ΔP	F_G	P'
67.66	Glenniesroth	0	0	123	114.23	65.50	2.16	1.0401	67.68
67.80	Dundee Man	4½	4.5	130	115.05	65.27	2.53	1.0470	67.45
67.61	Oscars List	nse	4.52	134	115.05	65.46	2.15	1.0398	67.65
67.07	Needless	1½	6.02	132	115.32	65.18	1.89	1.0350	67.36
68.20	Oui Oui	½	6.52	125	115.42	64.79	3.41	1.0636	66.96
67.66	Penguin Blue	nk	6.77	131	115.46	65.04	2.61	1.0486	67.22
68.89	Sailboat Ride	1¾	8.52	125	115.78	64.55	4.33	1.0812	66.71
66.67	Moon Stream	2	10.52	117	116.14	63.94	2.73	1.0517	66.08
68.15	Seeking	½	11.02	129	116.23	64.44	3.71	1.0695	66.60
68.00	Rhythm Stick	1¼	12.17	134	116.44	64.54	3.46	1.0647	66.70
							ave	1.0405	

In race 2, some of the horses who finished out of the first four, but within ten lengths of the winner, have high previous best performances. To include these in the sample would lead to a very highly increased rating being awarded to the winner, which is unlikely as these are seasoned handicappers. Instead it was decided that these horses had not run near to their best and so the first four to finish were taken as the sample.

Note that in the previous chapter the first six were selected as the sample, but as this race did not influence the final choice of F_G for the evaluation of time-based ratings, no change in the sample size was necessary.

Race 3. 4yo+ handicap, 9 runners, 16f (D=3218.56m), T_S = 210s, C_R' = 3.8155

P*	horse	btn	d	w	T	P	ΔP	F_G	P'
68.15	Whisky Chaser	0	0	134	225.60	63.18	4.97	1.0897	68.30
67.46	Birdman	¾	0.75	122	225.74	62.58	4.88	1.0890	67.64
67.61	Old Field	3	3.75	124	226.28	62.50	5.11	1.0933	67.56
68.20	Cheshire Cat	1½	5.25	136	226.55	62.97	5.23	1.0947	68.07
67.12	Zonguldak	2¾	8	118	227.05	61.98	5.14	1.0946	67.00
67.22	Three Dances	nk	8.25	122	227.10	62.15	5.07	1.0930	67.18
67.95	My Spirit	2¼	10.5	136	227.51	62.66	5.29	1.0961	67.75
67.51	Welsh Rock	2¼	12.75	127	227.92	62.12	5.39	1.0988	67.16
67.07	Concordial	32	44.75	116	233.74	59.86	7.21	1.1365	64.74

ave **1.0924**

In race 3 all runners finishing within ten lengths of the winner ran close to their previous best form, a value of F_G close to 1.09 being required to adjust their raw ratings. Therefore the average for this group gives reliable ratings for all runners in the race.

Race 4. 2yo seller, 10 runners, 6f (*D*=1206.96m), T_S = 74.22s, C_R' = 3.6481

P*	horse	btn	d	w	T	P	ΔP	F_G	P'
64.91	Marbella	0	0	118	74.22	63.86	1.04	1.0207	65.70
64.56	Open Sesame	8	8	118	75.67	62.38	2.18	1.0440	64.20
64.76	Glastonbury	½	8.5	123	75.77	62.52	2.24	1.0450	64.34
64.41	Alligator Boy	2¼	10.75	118	76.17	61.89	2.52	1.0512	63.69
63.78	Vermont Holiday	½	11.25	111	76.27	61.49	2.29	1.0467	63.28
64.46	Two Of Us	1¼	12.5	118	76.49	61.58	2.88	1.0586	63.38
65.00	Declaration	1	13.5	123	76.67	61.63	3.37	1.0685	63.43
64.36	Chessman	3¼	16.75	123	77.27	61.07	3.29	1.0673	62.86
63.87	Connelly Street	8	24.75	123	78.72	59.74	4.14	1.0858	61.50
-	Catkins	dht	24.75	118	78.72	59.52	-	-	61.28
							ave	1.0366	

In race 4 the winner finished well clear of the rest of the field. Using the winner and the two horses which finished within ten lengths gives a rating for the winner well ahead of its previous best.

Race 5. 3yo+ handicap, 13 runners, 5f (D=1005.8m), T_S = 57.8s, C_R' = 3.6347

P*	horse	btn	d	w	T	P	ΔP	F_G	P'
67.95	Victorious Nag	0	0	133	61.08	64.03	3.92	1.0795	68.46
67.95	Everest	1½	1.5	135	61.35	63.77	4.18	1.0850	68.19
68.20	Harlow Carr	shd	1.55	130	61.36	63.53	4.67	1.0953	67.94
68.49	Boardwalk	hd	1.65	133	61.38	63.65	4.85	1.0987	68.06
67.80	Steel Cutter	nk	1.9	133	61.43	63.59	4.22	1.0860	68.00
68.89	Himalayan Path	nse	1.92	146	61.43	64.18	4.71	1.0950	68.63
67.12	Narhdo	1¼	3.17	135	61.66	63.39	3.73	1.0762	67.79
67.80	Sun and Sand	1½	4.67	131	61.93	62.86	4.94	1.1016	67.24
68.30	Naajad	½	5.17	136	62.02	62.97	5.32	1.1091	67.36
67.66	Master Reynard	½	5.67	133	62.11	62.73	4.93	1.1014	67.10
68.00	Scimitar	½	6.17	121	62.20	62.07	5.93	1.1233	66.40
67.76	Bullfighter	2¾	8.92	133	62.70	62.00	5.75	1.1192	66.34
67.95	Spirit Level	½	9.42	130	62.79	61.76	6.19	1.1288	66.08
							ave	**1.0899**	

In race 5 all 13 runners finished within ten lengths of the winner, but it has to be assumed again that those who finished at the rear of the field ran below their best. Using the first six as the sample group gives a significant rating increase for the winner and runner-up, which is justified as this is a very competitive handicap.

Race 6. 4yo+ handicap, 8 runners, 8f (D=2413.92m), T_S = 152s, C_R' = 3.7248

P*	horse	btn	d	w	T	P	ΔP	F_G	P'
66.87	Chris D	0	0	119	158.20	64.81	2.07	1.0376	67.57
66.97	Bill Bones	7	7	117	159.47	64.10	2.87	1.0526	66.85
67.90	Maalesef	1¼	8.25	129	159.70	64.56	3.34	1.0609	67.33
67.46	Orange Canyon	2¾	11	127	160.20	64.23	3.23	1.0591	66.98
67.41	Red Sox Fan	2¾	13.75	122	160.70	63.76	3.65	1.0672	66.50
67.51	Golden Wolf	nk	14	132	160.75	64.21	3.30	1.0603	66.97
67.02	Woodside Ferry	3	17	114	161.29	63.12	3.90	1.0725	65.83
66.67	Baby Basket	1	18	120	161.47	63.31	3.36	1.0623	66.03

| | | | | | | | ave | 1.0504 | |

In race 6 there is again a clear group of three who meet the ten lengths criterion, and these are chosen as the sample group. This results in a significant increase in performance for the winner, compared with its previous best.

These form-based ratings P' are derived from a raw power rating, and then corrected, using the model of running expressed by the Power Equation. They are form-based rather than time-based, but nevertheless implicitly include all the dependencies of time on distance, weight, going, cost of running, and air resistance which are included in the model.

Note that the simple rules given here for sample group selection are sometimes broken when the ratings which result do not seem reasonable, based on all the information at hand. However it is much less difficult than it may at first appear, especially if the samples can be selected, and the results compared, using a few mouse clicks.

4.2 Conversion to pounds

Using the method described in section 3.5 the form-based power ratings can be converted to pounds and the results are shown in Table 4.1, which also shows the time-based ratings from Chapter 3.

		Power Rating (lb)	
race	horse	time-based	form-based
1	Felix	9	49
2	Glenniesroth	84	87
3	Whisky Chaser	37	100
4	Marbella	47	46
5	Victorious Nag	49	103
6	Chris D	71	85

Table 4.1 Comparison of time-based and form-based ratings in pounds

For races 2 and 4 (the two fastest races of the meetings) the form-based ratings are close to the time-based ratings. For the other races the form-based ratings are much higher. Based on form the winners of races 3 and 5 returned the best performances of the day.

4.3 Objective collateral form ratings

Evaluation of both types of ratings requires subjective judgments to be made when choosing the appropriate values of F_G to be used for the adjustment of raw power ratings. The same judgements are also required when calculating ratings in pounds by more conventional methods, for example when adjustments are made by a weight shift Δw.

Subjective judgements are more complicated for form-based ratings. As seen in the example meeting they require answers to questions such as (a) what is the best sample group to use as a reference? (b) is the rating which results for the winner reasonable? Useful information can be obtained from comments made by race readers and other issues connected to previous form. These can be difficult to assess accurately particularly early in the season when recent previous form may be scarce. Sometimes it becomes obvious that the judgement made about a particular race has not worked out well, and it is necessary to re-evaluate the decisions made about that race in the light of the way things turned out in future races.

When discussing the prospects of two horses A and B for a particular race, it is common for racing journalists to write something like ... "a month ago A beat C by two lengths at level weights. Subsequently B lost to C by a length while conceding 4lb. Therefore *on a line through C*, A should be 5lb better than B". If the writer was assuming that one length is equivalent to 3lb, the implication is that A is 6lb superior to B and that B is $(-3 + 4) = 1$lb superior to C, so A should be 5lb superior to B.

Such a discussion develops a so-called "form line", which may or may not work out in practice. It is, however, only one of millions of form lines which can interconnect all horses with all the other horses which run in a given period.

Form lines are also often invoked by stating that the form of a particular race has "worked out well". This usually means that the beaten horses have gone on to run well in subsequent races.

The following method evaluates collateral form ratings taking into account horses in all races which take place during the period, and their interactions with each other. It is objective in that adjustments of F_G or Δw are done automatically.

The method was developed by Murgatroyd (1975) and others for the adjustment of examination marks for classes of students where groups of them sit optional subjects, as often happens in the final year of a degree course. A high average mark in a particular subject may be because the group sitting that subject was above average, or may be because the examination paper in that subject was easy. By looking at all of the students' performances in all examinations, they can be graded according to ability, and the difficulty of the examinations can also be calculated. A similar objective method has also been used to rate sports such as athletics and soccer (Wilkins 1983).

As a first example consider 4 horses A, B, C and D which run in 5 races, and earn raw ratings (in pounds), as shown in the table below. Although the example is described here in terms of horse racing, the method is applicable to students sitting examination papers, or any competition for which a simple linear scoring system is used.

		1	2	3	4	5	all-race average, V_i
1	A	72		90		60	74.0
	B	60	83				71.5
i	C	50		77	83		70.0
n	D		40		84	65	63.0
R_j	raw race average	60.67	61.5	83.5	83.5	62.5	

(columns labelled $1 \ldots j \ldots m$)

Suffix i represents horse number ($i = 1 \ldots n$), while j represents race number ($j = 1 \ldots m$). The principle is that for each race, the race average R_j should be the same as the average of the all-race averages for the runners who ran in that race. Any discrepancy has to be removed by adding a shift Δw to the ratings of the race. This has to be repeated for all races. Since adding a shift alters the ratings in a race, it affects the all-race average, so the procedure needs to be applied iteratively until the solution converges.

Consider race 2 in the example. The average of the raw race ratings is (83+40)/2 = 61.5 (runners *B* and *D*). The average of the initial all-race ratings of runners *B* and *D* is (71.5+63.0)/2 = 67.25, which is 5.75 higher than the raw average for race 2. Therefore a shift of +5.75lb needs to added to the raw ratings of *B* and *D*. After the shifts have been calculated and added for all races, this completes one iteration.

The all-race averages then need to be recalculated and the process repeated iteratively until there is no change in the required race shifts. For the example, the calculations are shown below, where superscripts indicate the iteration number.

	1	2	3	4	5	V_i^0	V_i^1	...	Final
A	72		90		60	74.0	75.89	...	75.42
B	60	83				71.5	79.96	...	83.48
C	50		77	83		70.0	64.22	...	62.22
D		40		84	65	63.0	61.25	...	61.37
R_j^0	60.67	61.5	83.5	83.5	62.5				

$R_j^{\,1}$	71.83	67.25	72.00	66.50	68.50
Δw^1	11.16	5.75	-11.50	-17.00	6.00
$R_j^{\,2}$	73.36	70.65	70.05	62.75	68.57
Δw^2	12.69	9.10	-13.45	-20.77	6.07
... ...					
$R_j^{\,final}$	73.70	72.43	68.82	61.80	68.40
Δw^{final}	13.04	10.93	-14.68	-21.70	5.90

Illustrative calculations

$$V_1^{\,1} = (72 + 11.16 + 90 - 11.5 + 60 + 6.0)/3$$
$$= 75.89$$
$$R_3^{\,2} = (75.89 + 64.22)/2$$
$$= 70.05$$
$$\Delta w_3^{\,2} = 70.05 - 83.5$$
$$= -13.45$$

The average rating for each horse is given by the "final" column. Adding the final race shifts to the initial raw values gives the adjusted ratings shown below for each horse in each race.

	1	2	3	4	5
A	85.04		75.32		65.90
B	73.04	93.93			
C	63.04		62.32	61.30	
D		50.93		62.30	70.90

Although an intuitive approach has been used here to describe the method, Murgatroyd has shown that it produces optimal estimates of the performance of the competitors, based on the use of least-squares optimisation methods.

4.4 Sparsity-oriented programming

In the simple example above the results table had only 4 rows and 5 columns. If the method is applied to a full season of horse racing there will be thousands of horses running in thousands of races, so computer storage of the full table would require an array with many millions of entries. Even if such an amount of storage were available, computations based on the use of the full table would be extremely slow.

Fortunately, the table is *sparse* i.e. most of the entries are blank. It is only necessary to store the non-blank elements, which results in an enormous reduction in storage requirements. If there are say 4000 races to be analysed, and the average number of runners per race is 10, then only 40,000 elements need to be stored. In our simple example the elements taken row-wise could be stored as a linear array:

$$(72, 90, 60, 60, 83, 50, 77, 83, 40, 84, 65)$$

together with sets of integer pointer arrays which allow the start and end of each row and column in the original table to be located. This allows rapid scanning across rows and down columns for the calculation of averages, and the addition and deletion of rows and columns to the table.

Without these *sparsity-oriented* programming techniques, application of Murgatroyd's method to collateral form ratings would not be feasible. Further details can be found for example in the book by Pissanetzky (1984).

4.5 Application to power ratings

Murgatroyd's method needs to be modified so that it can be applied to the calculation of power ratings. Instead of a weight shift Δw, each race needs a multiplicative factor F_G to adjust the race ratings to match up with the all-race averages.

Based on the Power Equation the power rating for horse i in race j can be written

$$P_{ij} = F_{Gj}\, a_{ij} + b_{ij} \qquad (33)$$

where

$$a_{ij} = \frac{1}{F(T)}\left[\left(1+\frac{w}{W}\right)C_r'\frac{D}{T}\right] \; ; \; b_{ij} = \frac{1}{F(T)}\left[k\left(\frac{D}{T}\right)^3 + 2\left(1+\frac{w}{W}\right)\left(\frac{D}{T}\right)^2\cdot\frac{1}{T}\right]$$

a_{ij} and b_{ij} are calculated for each runner in each race from its time T and weight w.

Two elements a_{ij} and b_{ij} now need to be stored, rather than the single rating which was used in the previous example.

Instead of initial shifts, the initial values of F_G are set to 1.0 for all races. The all-race average ratings V_i are calculated by summing $(F_{Gj}a_{ij} + b_{ij})$ across the rows, and dividing by the number of elements in the rows. Then, as before, the average of the all-race averages for all runners who ran in race j is then calculated (Q_j).

The race average for race j is given by $(F_{Gj}\, a_{av} + b_{av})$ where a_{av} and b_{av} are the average values of these coefficients for race j. If this is equated to Q_j, we obtain the required change to F_{Gj}, as

$$F_{Gj} = \frac{Q_j - b_{av}}{a_{av}} \qquad (34)$$

For reference purposes, a detailed example of the adjustment of power ratings is given below, with the initial raw ratings shown first.

	1	2	3	4	5	V_i^0	V_i^1	...	Final
A	67.150		68.076		66.522	67.249	67.346	...	67.322
B	66.525	67.712				67.119	67.556	...	67.739
C	66.012		67.404	67.713		67.043	66.746	...	66.644
D		65.495		67.763	66.782	66.680	66.589	...	66.593

Q_j^0 67.137 66.899 67.146 66.862 66.965

(Note that in this case the Q_j shown are the all-race averages, not the column averages).

The initial raw ratings shown are the sum of the pairs of a and b calculated from each runner's performance, and are given below:

| race 1 || race 2 || race 3 || race 4 || race 5 ||
a	b	a	b	a	b	a	b	a	b
48.329	18.821			50.344	17.731			48.088	18.435
48.240	18.286	48.819	18.894						
47.979	18.033			49.852	17.553	48.440	19.274		
		47.962	17.533			48.860	18.903	48.499	18.283

The iterative adjustments then proceed as follows:

Q_j^1 67.137 66.899 67.146 66.862 66.965

F_G^1 1.0119 1.0061 0.9881 0.982 1.0065

Q_j^2 67.216   67.072   67.046   66.668   66.968

F_G^2 1.0136   1.0097   0.9861   0.978   1.0065

... ...

Q_j^{final} 67.235   67.166   66.983   66.619   66.958

F_G^{final} 1.0140   1.0116   0.9849   0.9770   1.0063

Applying the final values of F_G to the raw data gives the adjusted ratings as

	1	2	3	4	5	all-race average
A	67.825		67.315		66.826	67.322
B	67.198	68.280				67.739
C	66.682		66.651	66.598		66.644
D		66.053		66.639	67.089	66.593
Q_j final	67.235	67.166	66.983	66.619	66.958	

4.6　Large-scale application

When applying this method to the results of horse races, the entries in each column are the sample group selected. As a guide, the winner plus those finishing within 10 lengths should be used for flat racing, while for jumps an appropriate distance is 20 lengths. However this rule can be refined as desired. After the shifts Δw or correction factors F_G have been determined, these can then be applied to all the remaining runners who finished within a measurable distance.

Adding a new race adds a new column to the right-hand side of the table, and when the new ratings are calculated *all* figures in the table will be affected. Thus as the season progresses, the ratings in all races will change - every piece of information is used. After a few hundred races, the changes in the very early races will generally be quite small, but nevertheless there are changes. This is quite different from the conventional method, in which ratings published after a specific performance remain fixed thereafter.

The iterative procedure converges reliably, but rather slowly when there are several thousand runners, and it is essential to use an *acceleration factor* (Press *et al* 2007) to speed up the process.

However, the procedure will not converge at all if there are races in which no horse has ever raced against any of the other horses in the list. Collateral form, which compares the relative form of one horse relative to another, cannot be evaluated in these circumstances.

The table represents a interconnected network of links between runners, across rows and down columns. Before the iteration begins, one of the runners is selected (e.g. one which has had several runs) and then rows and columns are scanned to identify runners which are "connected". The scanning continues until no more can be found, and the remaining horses are marked as "disconnected" and excluded from the iterative solution process.

Large numbers of horses are disconnected during the first few weeks of a new season, until the collateral form ratings settle down. Starting off with form ratings brought forward from the previous season solves this problem, but obviously this cannot be done with ratings of two-year-olds.

4.7   Weight-for-age scales

While it is not common for time-based ratings to be corrected for age, it is normal to do so for form-based ratings.

One of the key changes which occurs as young horses grow is the increase in bodyweight W. A new-born foal weighs about 50kg. There is extensive published research on the subsequent growth rate, mostly concerned with the effects of different feeding methods. Figure 4.1 shows early growth results from six different sources. A study of 175 foals by Staniar *et al* (2004) showed a average mature weight of 542 kg, which was reached at the age of 7 years. This point is also shown on Figure 4.1. The growth is modelled very well as an exponential growth function, which is shown as a solid curve.

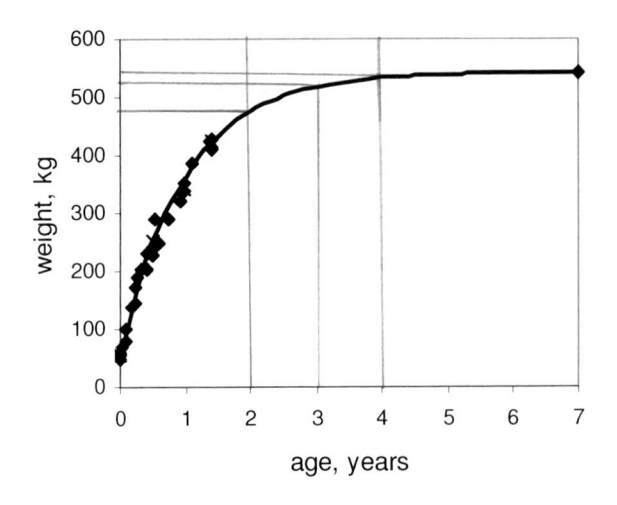

Figure 4.1 Growth in bodyweight of the thoroughbred

The majority of runners in flat races are aged from 2 to 4 years, and Figure 4.1 indicates that a value of 500kg (1100lb) is representative of this age range, and this was used for the model developed in Chapter 2. For jump racing horses are older, and a value of 540kg (1200lb) was used, which is just less than Staniar's fully mature value.

A. Effect of W on ratings

A bodyweight of 1100lb has been used for the calculation of all power ratings on the flat using the Power Equation. In British racing the weights of the horses are not measured and published before a race, so there is no option but to use an assumed standard value.

This leads to an error in the calculated rating. *All other model parameters remaining constant*, a heavier horse should be able to run a race more quickly than a lighter horse, mainly because the weight carried will have less effect. The bodyweight W only occurs in the Power Equation in the ratio w/W, so the effect of an increase in W has the same effect as a corresponding decrease in w.

Consider a standard horse with a bodyweight of 1100lb and another horse with identical physiological characteristics but which weighs 1150lb. The times taken to run distances from 5f to 16f, obtained from the Power Equation are shown in columns 2 and 3 of Table 4.2. Column 3 gives the time differences between the two horses, and in column 4 this is converted to pounds.

Now consider a race between these two horses, both carrying the standard weight w_S, which is run in the standard time and which results in a dead heat. In the absence of any information about their bodyweight, both horses would normally be awarded the standard rating of, say, 100. However, the heavier horse should have done better than this, and its true rating should be 100 - 5.5 = 94.5.

So any method of calculating ratings which does not take account of bodyweights will contain inbuilt errors - true ratings for heavier-than-standard horses will be overestimated, while for lighter horses they will be underestimated.

D, f	T_{1100}, s	T_{1150}, s	$\Delta T, s$	$\Delta w, lb$
5	58.68	58.50	-0.183	-5.5
6	71.57	71.34	-0.229	-5.5
7	84.73	84.45	-0.275	-5.5
8	98.05	97.73	-0.321	-5.5
10	124.98	124.57	-0.413	-5.5
12	152.09	151.59	-0.504	-5.5
14	179.28	178.69	-0.594	5.5
16	206.52	205.84	-0.685	-5.5

Table 4.2 Effect of bodyweight 50lb above standard

Fortunately this error, although possibly quite large, is generally self-cancelling.

The main purpose of handicapping is to evaluate the effect of weight carried for *future* races. If the two horses discussed were to meet again in the future, again carrying standard weight, they would both receive a rating of 100 if bodyweight is ignored. If bodyweight were taken into account, the standard 1100lb horse would be given a rating of 100. The heavier horse would have a raw rating of 94.5, but because of its bodyweight advantage, this would need to be increased by 5.5lb, bringing its rating back to 100.

So errors introduced by differences in bodyweight *alone* can be ignored, because of the self-cancelling effect when future races are considered.

B. Weight-for-age

A problem arises, however, if the weight of one of these horses increases due to increased growth. Assuming that all other parameters remain constant, this would result in an increase in potential performance. The number of pounds to be added back to its rating would be greater than any previous underestimate. This is the reason for the use of weight-age-scales; they give the number of pounds a horse is short of maturity for a given age and distance to be run, so that ratings can be adjusted accordingly.

If bodyweight were the only physical attribute to change as the horse grows, the required weight allowance could be easily calculated using the Power Equation and Figure 4.1. However, the model parameters E_0, T_E, P_m, T_P will also change, and little information is available on how they change as a horse grows. $\dot{V}O_2max$, which is related to P_m, is slightly lower for 2-year-olds than for older horses, and for 1-year-olds it is significantly less (Rose *et al* 1990), but it seems likely that the ability to convert this to speed across the ground is much reduced for younger horses. Development of cardiovascular systems, muscle efficiency, running gait and exhaustions limits with maturity are important, and will also affect the model parameters. There is insufficient data from exercise physiology testing which would enable the metabolic model to be extended to include these effects.

The metabolic model can be adjusted to fit the official or other weight-for-age scales. However, this approach is not compatible with the key idea of this book, which is to develop models from first principles, based on known data from exercise physiology.

In the absence of a metabolic model for WFA, the Power Equation can be used as it stands for form-based power ratings as well as time-based power ratings. This means that a horse may show a sudden improvement in its rating if it has not run for some time, and it has matured well. Of course the opposite is also true. Many horses who show great promise as 2-year-olds fail to fulfil that promise as they grow into 3-year-olds. WFA scales assume that all horses will improve equally as they grow, and adjusts their ratings accordingly. Without WFA scales, horses are awarded ratings which depend on what they have actually achieved, rather than what they promise.

Chapter 5. Allowance for Weight in Future Races

One of the main reasons for calculating racehorse ratings is to be able to assess their chances of winning future races - usually but not always for betting purposes. Conventional ratings expressed in pounds are usually corrected according to the weight to be carried, as illustrated in Table 5.1.

Horse	Current rating lb	Weight st lb	Corrected rating lb
A	86	9 - 2	84
B	77	9 - 0	77
C	81	8 - 7	88

Table 5.1 Correction of rating to weight carried in "today's" race

Column 2 gives the current rating of the horses. If all 3 horses carried the standard weight of 9-0 then horse A would have the best chance of winning (when judged by weight alone). However the weights to be carried are different. Horse A is set to carry 2lb more than standard and its rating needs to be decreased by 2. When this adjustment is done for all three horses, horse C has the "best chance at the weights" and would be the *selection* in this race, according to the ratings.

However there are many other factors which affect the result of a race, and which are very important in determining the outcome. Just a few of these are listed below.

- Fitness - the horse can only run to its best previous rating if it is brought fully fit to the track by its trainer. (See section 2.1B). Good form in recent races can be a guide to this.

- The rider - horses do not race on their own, and the skill of the rider can make the crucial difference, with respect to general riding skills, but particularly with respect to the judgement of pace and the tactics employed. (See section 2.5E).

- The race distance. (See section 2.5F).

- The horse may have particular preferences - some horses exhibit preferences for certain types of going, right-hand or left-hand tracks, particular racecourses, ways of being ridden, and so on.

- The draw - at certain flat racing tracks this can be a significant advantage or disadvantage.

- Consistency - a horse which has shown consistency will obviously be more likely to be able to run to its previous best.

- Class of opposition.

- The type and class of the race.

- Luck - even if everything is perfect, a bad break in running can ruin a horse's chances.

Detailed discussions of these points, and many other important influences, fill the pages of the daily and weekly racing publications. Some of the non-weight factors can be quantified in some way and used in predictive models which attempt to calculate the probability of winning for each runner in a race.

There are so many factors other than weight carried, which affect the outcome of races, that sometimes it is suggested that weight carried should be ignored altogether. However, the top-rated horse wins much less frequently in handicaps than in non-handicap races. This indicates that the allocation, by the official handicapper, of weights to be carried in handicaps, does indeed have a clear effect. Horses are not machines, and it is unrealistic to expect all handicap races to end in blanket finishes, but anyone who has compiled their own private handicap will testify to the fact that many seasoned handicappers run close to their rating time after time.

Dick Whitford, the distinguished handicapper, once wrote "*more horses will win races because they are favoured by the weights than for any other reason*".

When using power ratings, it is not possible to find which horse is "best in" at the weights by simply adding or subtracting pounds according to the weight to be carried. For each horse the power rating *P* and weight to be carried *w* are known, and so the Power Equation can be solved to find the time *T*. This is the time the horse would take to run the race if it performed according to its rating, and takes account of the actual weight to be carried. The horse with the lowest value of *T* will be the selection, based on the weights. For this calculation F_G can be taken as 1.0, as the relative rankings of the horses in a future race is not altered by changing F_G.

As an example, consider an attempt to predict the result of race 6, over 8f, which was used as an example in the previous chapters. Columns 1 to 4 of Table 5.2 show the horse's name, previous rating, and weight to be carried in stones and pounds.

Column 5 shows the time predicted by the Power Equation, assuming that each horse ran to its rating, and using standard going conditions. According to this, Woodside Ferry, with the lowest time, is the selection, while Golden Wolf is "worst in" at the weights.

horse	P	st	lb	T, s	result
Woodside Ferry	67.02	8	2	150.90	
Bill Bones	66.97	8	5	151.17	**2nd**
Chris D	66.87	8	7	151.36	**1st**
Baby Basket	66.68	8	8	151.45	
Red Sox Fan	67.41	8	10	151.63	
Orange Canyon	67.46	9	1	152.09	
Maalesef	67.90	9	3	152.28	**3rd**
Golden Wolf	67.51	9	6	152.55	

Table 5.2 Use of Power Equation to predict race times

The actual result of the race (see note in the first paragraph of section 3.4) is shown in the final column. Woodside Ferry finished out of the first three, while the second and third-rated horses finished second and first respectively. Use of the ratings alone did not predict the winner in this case, but the result is reasonably satisfactory - obviously some combination of the non-weight factors discussed above also contributed to this result.

This book is not primarily concerned with the way in which ratings are used, only with the way they are calculated and their relationship to a bioenergetic model of running. However some further information on the use of ratings for betting is included in Appendix B.

Chapter 6. Conclusion

A whole-body metabolic model of the running of racehorses has been described, which balances the anaerobic and aerobic sources of energy developed by the horse with the energy needed for initial acceleration, to overcome air resistance, and to sustain running speed. The model parameters have been calculated by fitting to standard time data for flat and jump racing at British racecourses. The interactions between weight carried, time, distance, going, and other factors have been investigated.

Application of the model to the calculation of racehorse ratings leads to the use of metabolic power as a measure of performance, and some basic examples have been given showing how this can be used for evaluation of both time-based and form-based ratings.

A brief discussion of the non-weight factors which affect performance, and the usage of ratings has also been included.

Further research is needed on the exercise physiology of immature horses, and how this develops with growth, before the model can extended to represent this development from first principles.

References

This list only includes the key publications used in the development of the models in this book. Further information on biomechanics and exercise physiology can be found in Hay (1993), and McCardle et al (2005), while Morton (2006) gives a review over 120 papers in the field of whole-body metabolic models (for humans only).

Busso, T. and Chatagnon, M. Modelling of aerobic and anaerobic energy production in middle-distance running. *Eur J Appl Physiol 97*, pp745-754, (2006).

Cavagna, G. A., Komarek, L. and Mazzoleni, S. The mechanics of sprint running. *J. Physiol. London. 217*, pp709-721, (1971).

Cavagna, G.A., Saibene, F.P. and Margaria, R. Mechanical work in running. *J Appl Physiol 19*, pp249-256. (1964)

Davies, C. T. M. Effects of wind assistance and resistance on the forward motion of a runner. *J. Appl. Physiol. 48*, pp702-709, (1980).

di Prampero P.E, Capelli C, Pagliaro P, Antonutto G, Girardis M, Zamparo P, Soule R.G. Energetics of best performances in middle-distance running. *J. Appl. Physiol. 74*, pp2318–2324. (1993).

di Prampero P.E, Fusi S, Sepulcri L, Morin J.B, Belli A, Antonutto G. Sprint running: a new energetic approach. *J. Exp. Biol. 208*, pp2809–2816, (2005).

Dickinson, D. How to compile your own handicap. *Raceform Ltd.* (1997).

Duffield, R. and Dawson, B. Energy system contribution in track running. *New Studies in Athletics. IAAF*, pp47-56, (2003).

Epstein, R.A. The theory of gambling and statistical logic. *Academic Press*, New York. (1967).

Hay, J.G. The biomechanics of sports techniques. *4th ed, Prentice-Hall*, (1993).

Hill, A.V. The air resistance to a runner. *Proc. Roy. Soc. B*. pp380-385, (1927).

Keller, J.B. A theory of competitive running. *Physics Today 26*, pp42-47, (1973).

Kelly, J.L. A new interpretation of information rate. *Bell Syst Tech J 35*, pp917-926. (1956).

Lloyd, B. B. The energetics of running: an analysis of world records. *Adv. Sci. 22*, pp515-530, (1966).

McArdle, W.D., Katch, F.I. and Katch, V.L. Essentials of exercise physiology. *3rd ed. Lippincott Williams and Wilkins*, (2005).

McCutcheon, L.J., Geor, R.J., and Hinchcliff, K.W. Effects of Prior Exercise on Muscle Metabolism During Sprint Exercise in Horses. *The American Physiological Society*, pp1914-1922 (1999).

Morton, R.H. The critical power and related whole-body bioenergetic models. *Eur J Appl Physiol 96*, pp339-354, (2006).

Mureika, J. A Realistic Quasi-Physical Model of the 100 Metre Dash, *Canadian Journal of Physics 79*, pp697-713, (2001).

Mureika, J. A Simple Model for Predicting Sprint Race Times Accounting for Energy Loss on the Curve, *Canadian Journal of Physics 75*, pp837-851, (1997).

Murgatroyd, P.N. An adjustment procedure for examination marks in optional subjects. *Int J Math Educ Sci Technol 6*, 435-444, (1975)

Peronnet, F., and Thibault G.. Mathematical analysis of running performance and world running records. *J. Appl. Physiol. 67*, pp453-465, (1989).

Pissanetsky, S. Sparse Matrix Technology. *Academic Press,* London, (1984).

Potard, U.S.B., Leith, D.E., and Fedde M.R.. Force, speed, and oxygen consumption in thoroughbred and draft horses. *The American Physiological Society,* pp2052-2059, (1998).

Press, W.H., Teukolsky, W.T., Vetterling, W.T. and Flannery, B.P. Numerical Recipes: The Art of Scientific Computing. *3rd ed. Cambridge University Press.* (2007).

Pugh L. G. C. E. The influence of wind resistance in running and walking and the mechanical efficiency of work against horizontal or vertical forces. *J. Physiol. London 213,* pp255-276, (1971).

Roberts, T.J., Kram, R., Weyand, P.G., and Taylor, R. Energetics of bipedal running. I. Metabolic cost of generating force. *J ExpBiol 201,* pp2745–2751, (1998).

Rose, R.J., Hendrickson, D.K. and Knight, P.K. Clinical exercise testing in the normal Thoroughbred horse. *Australian Veterinary Journal,* vol 67, pp 345-348, (1990).

Staniar, W.B., Kronfeld, D.S., Treiber, K.H., Splan, R.K. and Harris, P.A. Growth rate consists of baseline and systematic deviation components in Thoroughbreds. *J. Anim. Sci.* vol 82, pp 1007-1015. (2004).

Wilkins, R. Electrical networks and sports competition *Electronics and Power, IEE,* pp 414-418, (1983).

Appendix A - a note on betting

This section is included for those who may have opened this book in the hope that improved methods of calculating ratings will enable them to make a fortune by betting on horses.

Consider a series of N bets struck with a unit stake at odds against of s (i.e if the odds are 7-2, $s = 3.5$ and so on). If N_w of these bets are successful, the winnings will be $N_w s$ units. The remaining bets will be losing bets, with a loss of $(N - N_w)$ units. The net profit over the series will be

$$E = N_w s - (N - N_w) \qquad (A.1)$$

Assuming that N is large and dividing by N gives

$$e = ps - (1 - p) \qquad (A.2)$$

where e is the profit per unit stake, and p is the probability of winning. Putting $e = 0$ and solving for p gives the break-even criterion

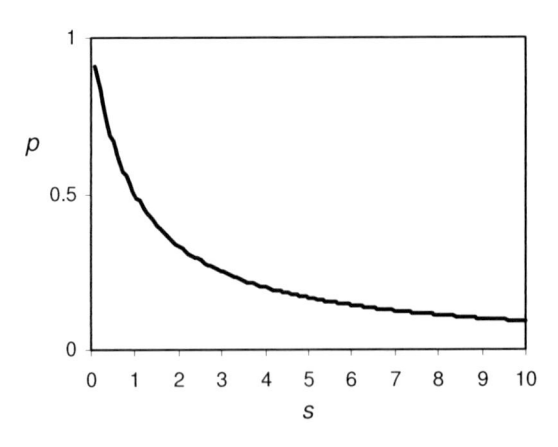

$$p = \frac{1}{s+1} \qquad (A.3)$$

Figure A.1 Break-even probability as a function of price taken

For even chances ($s = 1$) the win probability needs to be greater than 1/2 to make a profit, while if $s=3$ the probability needs to be greater than 1/4 and so on. The relationship between s and p is shown in Fig.A1.

Consider first the casino game of roulette. On the European roulette wheel there are the numbers 1-36, and a winning bet on any one of these numbers pays out at odds of 35-1. These would be fair odds if there were only these 36 numbers on the wheel, but there is also a number 0, which is a losing bet, and so the win probability of a bet on any of the numbers 1 to 36 is 1/37. The expected profit for a long series of bets is then given by equation (A.3), as

$$e = \frac{1}{37}(35) - \left(1 - \frac{1}{37}\right) = -0.027 \quad (A.4)$$

The expected profit is negative: on average the punter will lose 2.7% of his stake *per bet*.

A variety of other bets are available at roulette, but they all have a built-in loss of 2.7% or more per bet.

While individual bets will result in wins as well as losses, *in the long run* the casino will win. Many fancy staking plans exist for roulette players, but they cannot change the built-in advantage for the casino. Although it is possible to "get lucky" with a winning streak, financial ruin is inevitable *if the player continues to play* (Epstein 1967).

For the average punter there is a further, hidden, disadvantage. If two players bet either heads or tails on the flip of a fair coin, it would appear intuitively that after a very large number of bets the players would finish up all square.

Not so. During the game each player will encounter at some time a long run of consecutive losses, requiring each to dip into his capital resources, and the player with the larger capital resources is likely to be able to withstand this better. In a fair game, financial ruin is inevitable in the long run when playing against an "infinitely rich adversary" to use Epstein's phrase. For the average punter, bookmakers and casino owners fit the description of "infinitely rich adversary" rather well.

So in betting there is a built-in bias in favour of rich layers as well as any advantage they may have due to the odds offered. Despite this most punters continue to believe their luck will change, sometimes invoking a non-existent "Law of Averages", hoping that after a long losing run, their chances of winning will increase. In fact each bet is an independent event. If 10 flips of a coin gives 10 heads in succession, then the probability of another head on the 11th flip is still 1/2.

For racing the nearest equivalent to a spin of the roulette wheel is to try to pick the winner of a race by choosing a horse at random, similar to closing ones eyes and sticking a pin into the daily newspaper.

Results obtained using this method are shown in Table A.1 below. These, (and the other results which follow) were obtained for two consecutive complete seasons of National Hunt racing (almost 7000 races of all types). A random number generator was used to make the selection. In the table $p = N_w/N$, s is the average starting price of the winners, and LLR is the longest run of consecutive losing bets encountered.

p	s	e	LLR
0.133	4.205	-0.307	49

Table A.1 Results of backing a random selection

The situation is very, very much worse than roulette, with expected losses of 30.7% per unit stake. These losses vary depending on how the random numbers are generated. Very few races are won with a starting price of 100-1, but this method of selection can occasionally find such a winner, which changes the overall picture a little, but expected losses within the range of 30-40% are typical. With a longest losing run of 49, any roulette-style staking plan which involved increasing stakes after a loss would be catastrophic.

However, racing is different from roulette in that the profitability can be improved by sensible betting - picking a horse at random is not a good idea! An improvement can immediately be obtained by backing the favourite. The results in Table A.2 would be obtained by backing the clear favourite at starting price for the same two seasons of jump racing.

p	s	e	LLR
0.352	1.437	-0.142	22

Table A.2 Results of backing the favourite

The win percentage has increased considerably, from 13.3% to 35.2%, but the large reduction in the average starting prices means that the expected profit is still negative. Although the losses have been cut drastically, to 14.2%, the method is still much worse than roulette.

Table A.3 shows results obtained using ratings calculated by the objective collateral form method described in Chapter 4 and using times predicted by the Power Equation to make the selection. To obtain these results the ratings for hurdles and chases were computed separately, starting from scratch at the beginning of the first season.

p	s	e	LLR
0.218	3.062	-0.114	24

Table A.3 Results using objective collateral form ratings

The win percentage is now about 22%, but the improved average starting prices mean that the expected losses are reduced to 11.4%

These results show that while ratings are a sound starting point for finding winners, the uncritical application of them to all races regardless of other factors is foolish and will still be far from profitable. They need to be used in conjunction with the many other factors which influence the outcome of a race, discussed in Chapter 5 to assess the actual probability of winning, compared with the returns promised by the available price. A bet should only be struck if it offers value, i.e.

$$p > \frac{1}{s+1} \quad (A.5)$$

It is easy to increase the win probability, for example only backing a horse with a top jockey, or only if it won its last race, but these simple factors are so obvious that the starting prices obtained fall dramatically, giving no improvement in profitability. The requirement of (A.5) applies equally for all types of betting, such as pool betting.

Meeting the value requirement of (A.5) is a formidable challenge, but if such a favourable method of betting can be found ($e>0$) it is best to adopt a "prudent" staking strategy, i.e. increase stakes after a winner and reduce them after a loser. For optimum growth of capital, the stake should be a fraction f of the available capital, where $f = e/s$. So if $e = +0.1$ and $s = 2.0$ the stake should be 5% of the capital. This result is derived from the work of Kelly (1956), who first deduced an optimal strategy for betting in favourable games.

If the results turn out to be unfavourable, the capital runs down slowly, and so the fractional staking method is also very suitable for "betting for fun", in which the punter starts with an initial capital, which he is prepared to lose, but which he *hopes* might increase.

Appendix B - numerical methods

A high degree of numerical precision is required for computation of many of the values in this book. This may seem unnecessary at first sight, as conventional racehorse ratings are numbers like 100, rounded to the nearest pound. However the ratings computed by the Power Equation are spread about a relatively narrow zone around the standard power P_m. Their calculation, and in particular the solution for T using bisection requires high numerical precision. For all calculations in this book, a 64-bit representation of floating-point numbers was used.

However, display of the results to several decimal places would make the text unreadable, so they have been rounded throughout to obtain a good compromise between precision and readability.

Appendix C - Dickinson's model

Dickinson (1997) suggested that the relationship between weight and distance beaten can be obtained by simply dividing the weight of the horse plus rider by the race distance D. For a 5 furlong race, D is 1005.8m or 372.5 lengths (if 1 length = 2.7m). If the horse plus rider weigh 1226lb and we divide by 372.5 we get 3.29lb/length, which is quite close to the relationships commonly used.

This is an astonishing result, as there is no physical basis for the calculation as described and therefore no reason why it should apply to horses rather than hot-rod cars or artillery shells. Adding 3lb to the weight of an 1100lb artillery shell fired over a distance of 1005.8m is unlikely to slow it down by the length of a horse.

Yet the result is so close to the right answer that it is worth examining what kind of model could produce this result.

We neglect the effect of the start and assume that the race is run at a constant speed. and that this speed is governed by a forward propulsive force, which increases with the weight of the horse W, and a resistive drag force which increases with $(W + w)$ and increases linearly with speed. These are reasonable assumptions. If the speed is constant, Newton's Laws require that

$$k_1 W = k_2 (W + w) \frac{D}{T} \qquad (C.1)$$

where k_1 and k_2 are unknown constants. This then gives

$$D = \alpha \frac{W}{W + w} T \qquad (C.2)$$

where $\alpha = k_1 / k_2$.

Solving for T we obtain

$$T = \frac{W+w}{\alpha W} D \qquad (C.3)$$

Differentiating, a small change in weight carried Δw will result in a change in time given by

$$\Delta T = \frac{D}{\alpha W} \Delta w \qquad (C.4)$$

Using (C.2) we then obtain the change in distance travelled, as

$$\Delta D = \alpha \frac{W}{W+w} \cdot \frac{D}{\alpha W} \Delta w$$

giving

$$\frac{\Delta w}{\Delta D} = \frac{W+w}{D}$$

which is the desired result.

Appendix D - list of principal symbols

$\dot{V}O_2$ = rate of oxygen absorption, $mlO_2/min/kg$

$\dot{V}O_2 max$ = maximum value of $\dot{V}O_2$

A = projected area for air resistance, m^2

a = see equation 33

b = see equation 33

C_D = drag coefficient

C_R = energy cost of running, J/kg/m

C_R' = energy cost of running at a specific track and distance, J/kg/m

D = race distance, m or f

d = distance beaten, lengths

e = 2.71828 ...

E' = energy available from anaerobic sources, J/kg

E'' = energy available from aerobic sources, J/kg

E_0 = initial value of anaerobic store, J/kg

E_A = energy needed to overcome air resistance, J/kg

E_K = kinetic energy, J/kg

E_R = energy cost of running a distance D, J/kg

F = viscous drag force, N

f = viscous drag force, N/kg

F_G = going coefficient

g = acceleration due to gravity, m/s^2

H = change in altitude, m

k = air resistance coefficient, $Ws^3/kg\text{-}m^3$

m = mass, kg

p = instantaneous power, W/kg

P = average power over duration T, W/kg

P' = corrected power rating, W/kg

P^* = previous best power rating, W/kg

P_A = power needed to overcome air resistance, W/kg

P_K − power needed to supply kinetic energy, W/kg

P_m = maximum power, W/kg

P_R = power needed to supply energy cost of running, W/kg

q = coefficient to convert from power ratings to pounds

R_j = pounds shift needed for race j

Q_j = all-race average power rating, W/kg

SPF = seconds per furlong

SSE = sum of squares of errors

T − duration of race, s

t = instantaneous time, s

$T_0 = E_0/P_m$, s

T_E = time constant for decay of anaerobic stores, s

T_P = time constant for growth of aerobic power, s

T_S = standard time, s

V = average speed, m/s

V_F = finishing speed, lengths/s

V_i = all-race average rating

V_W = wind speed, m/s

w = weight carried, lb

W = weight of the horse, lb

w_{REF} = reference value for ratings in pounds, lb

w_S = standard weight, = 126lb (flat) or 168lb (jumps)

Δw_G = going correction, lb

ρ = air density, kg/m^3

Δ represents a small change in the quantity that follows

Σ represents the sum of the items that follow